RUWENZORI

THE TRANSLATION FROM THE ITALIAN HAS BEEN MADE BY
CAROLINE DE FILIPPI, *née* FITZGERALD.

———— ————

THE ILLUSTRATIONS ARE FROM PHOTOGRAPHS TAKEN BY
VITTORIO SELLA, MEMBER OF THE EXPEDITION.

SUNSET ON VICTORIA NYANZA
FROM THE PAINTING BY A. FITZGERALD

RUWENZORI

AN ACCOUNT OF THE EXPEDITION

OF

H.R.H.
PRINCE LUIGI AMEDEO OF SAVOY
DUKE OF THE ABRUZZI

BY

FILIPPO DE FILIPPI. F.R.G.S.

WITH A PREFACE BY

H.R.H.
THE DUKE OF THE ABRUZZI

SECOND IMPRESSION

GREENWOOD PRESS, PUBLISHERS
NEW YORK

In the spelling of the native names the usage established by the Royal
Geographical Society has been followed.

Originally published in 1909
by Archibald Constable and Company, Ltd.

First Greenwood Reprinting, 1969

Library of Congress Catalogue Card Number 68-55185

SBN 8371-1349-0

PRINTED IN UNITED STATES OF AMERICA

Dedicated

BY PERMISSION

TO HER MAJESTY

THE QUEEN DOWAGER OF ITALY

MARGHERITA DI SAVOIA.

PREFACE.

THIS book contains a detailed account of the expedition which I led from April to September of 1906, to explore the snow range of Ruwenzori, upon the borders of Congo and Uganda, in the centre of Equatorial Africa.

The book includes the data of observations, and all the facts upon which the geographical and scientific results of the expedition are based. These results I have already given in outline in my lectures before the Italian Geographical Society and the Royal Geographical Society, which were subsequently published in the "Bollettino" and "The Geographical Journal."

I had not at my disposal the time requisite for writing myself a full account of this journey. My companions were, for various reasons, equally unable to undertake the task. Cav. Filippo De Filippi had already published an accurate and painstaking account of a previous exploration, in which he had been one of my companions. It, therefore, occurred to me to request him to write the story of the Ruwenzori Expedition from our notes and journals.

Preface.

The task was difficult, even with the help of Cav. Uff. Vittorio Sella's splendid photographs, which, in a measure, filled out the bare outlines of our diaries. I, therefore, take this opportunity of expressing my deep sense of obligation to Cav. Filippo De Filippi, and of recording the pleasure given me by his acceptance of my proposal.

To this volume, which contains a narrative of the journey and of the actual exploration work of the expedition, together with the meteorological and astronomical notes, will be added a second volume,* containing the geological and mineralogical report of Dr. A. Roccati, together with reports upon the zoological and botanical specimens brought back by us.

I must here tender my thanks to all the distinguished men of science who have collaborated in the study and illustration of our collections.

Luigi di Savoia

ROME,
April, 1908.

---

* Published in Italian only—U. Hoepli, Milan.

CONTENTS.

Contents.

ILLUSTRATIONS.

Illustrations.

Illustrations.

Illustrations.

Illustrations.

LIST OF PLATES.

List of Plates.

CHAPTER I.

THE DISCOVERY AND PREVIOUS EXPLORATIONS OF RUWENZORI.

Stanley's first sight of the Snow-peaks—Ruwenzori and the "Mountains of the Moon" of Ptolemy—Discussions and Dissensions between Geographers— Exploration of Stairs, Stuhlmann and Scott Elliot—Moore discovers the Glaciers at the head of the Mobuku Valley—Repeated attempts to climb the Mountains from the Mobuku Valley—Ascent made by David upon the western slope—British Museum Expedition—First peaks ascended—What was known about Ruwenzori in the beginning of 1906.

ON the 24th of May, 1888, Henry Stanley, in the midst of his longest and most venturesome African journey, while crossing the narrow strip of coast which forms the south-west bank of Lake Albert Nyanza, between Nsabé and Badzwa, beheld for the first time the snowy peaks of Ruwenzori: "When about five miles from Nsabé camp, while looking to the south-east and meditating upon the events of the last month, my eyes were attracted by a boy to a mountain, said to be covered with salt, and I saw a peculiar-shaped cloud of a most beautiful silver colour which assumed the proportions and appearance of a

1

Chapter I.

vast mountain covered with snow. Following its form downward, I became struck with the deep blue-black colour of its base, and wondered if it portended another tornado ; then, as the sight descended to the gap between the eastern and western plateaux, I became for the first time conscious that what I gazed upon was not the image or semblance of a vast mountain, but the solid substance of a real one, with its summit covered with snow."

" Ruwenzori " is the one among many native names by which, in Stanley's opinion, the mountain is most widely known in the surrounding region.

Of all the explorers who in the preceding twenty years had travelled through these regions and sailed upon the waters of the lakes at the foot of the chain, not one had suspected the near presence of vast tracts of eternal ice and snow hidden from all eyes in the impenetrable cloak of cloud and mist.

In 1864, Sir Samuel Baker had given the name of " Blue Mountains " to the vast shapes faintly seen looming through the mists of the plain to the south of the Albert Nyanza. He did not, however, form any adequate conception of their real proportions.

Stanley himself, in the December of 1875, when actually encamped upon the eastern slopes of the chain, relates, but without comment, the descriptions given by the natives of the shining white colour and intense cold of peaks which he could not see but which were said to be towering above him.

Sir Harry Johnston mentions certain private letters written in 1876 by Romolo Gessi during his first complete exploration of the shores of the Albert Nyanza. In these letters mention is made of a strange vision which the writer saw in the sky, as if of mountains covered with snow. Possibly he ascribed this vision to an hallucination. The fact remains that the discovery of Ruwenzori was reserved neither for him, nor for Emin Pasha, or Mason, both of whom subsequently visited the lake.

TUSKS CONFISCATED BY THE GOVERNMENT, UGANDA.

Stanley is probably right in attributing the extraordinary lack of atmospheric transparency, which renders these mountains invisible even in fair weather, to vapours exhaled from the surrounding plains and exposed to the heat of the tropical sun. Occasionally a breeze sweeps away these vapours. At such times, as if by magic, the snow-clad ranges loom into sight only to vanish again and leave the onlooker in doubt and uncertainty as to the actual reality of the magnificent vision vouchsafed to him.

The discovery of this vast system of snow mountains shedding their waters into lakes, whence one of the greatest Nile branches takes its origin, finally brought to an end that quest after the sources of the classic river which had played so large a part in the history of geographical investigation.

Chapter I.

After a lapse of twenty-four centuries the verse of Aeschylus—
" Egypt nurtured by the snow "—once more receives its literal
signification. The " Mountain of Silver " (ἀργυροῦν ὄρος), the
source of the Nile, according to Aristotle, is at last revealed.
Strange indeed are the vicissitudes of human knowledge.

This classical tradition of the Nile issuing from vast lakes
fed by snowy mountains was tenaciously preserved through
subsequent history. We find it repeated in the description
of the " Mountains of the Moon " taken by Ptolemy, with
modifications of his own, from the writings of Marinus of Tyre.
The same story recurs in the writings and maps of Arab
geographers in the Middle Ages; and again in the works of
Western European compilers, such as the Prior of Neuville
les Dames and Alphonse de Saintorge. In fact, notwithstanding the absolute lack of any confirmation of their
existence, these mountains and lakes, indicated with uncertain
forms and doubtful and varying geographical situation, never
wholly disappeared from our maps of Africa up to the time
of their actual discovery.

The belief in snow-clad mountains at the sources of the Nile
had persisted with peculiar tenacity among the natives of the
East Coast. Possibly it received fresh confirmation from time
to time through news gathered from the caravans which brought
ivory and slaves from the interior. Burton, Speke and Baker
heard it again and again, and with positive affirmation, both
from the Arabs and from the natives of Zanzibar.

The discovery of Mt. Kenya and Mt. Kilimandjaro by the
German missionaries Krapf and Rebmann in 1848–49 seemed
for the time to settle the question. These mountains, however, are connected neither with the lakes nor with the Nile.
In 1861 Speke believed that he had discovered the " Mountains

of the Moon" in the volcanic chain which stretches between Lake Kivu and Lake Albert Edward, more especially in the highest of the volcanic peaks, Mt. Nfumbiro; but

NATIVE HUT IN UGANDA.

none of these mountains are covered with snow. Stanley had certainly far stronger grounds for his opinion that the "Mountains of the Moon" of Ptolemy are to be identified with Ruwenzori, which alone answers in all essential points to the descriptions of the ancient geographers. It consists of a vast mountain range covered with everlasting ice and snow and shedding its waters into the basin of the Upper Nile from all its slopes. Rising, as it does, out of the midst of a tropical

Chapter I.

landscape, it forms a spectacle at once so imposing and so un-expected as to strike the imagination of those who behold it more forcibly than any other feature of the whole region, and so impresses itself upon their memory as not to be effaced by any subsequent vicissitude or experience of their journey.

The opinion of Stanley, however, met with numerous opponents, including a number of competent geographers.

The German explorer, Dr. O. Baumann, discovered the sources of the Kagera, the greatest tributary of the Victoria Nyanza, in the mountains of Missóssi ya Mwesi, in Urundi, a district situated to the north-east of Lake Tanganika. These he considered to be the mountains mentioned by Ptolemy; Missóssi ya Mwesi does, as a matter of fact, mean literally "Mountains of the Moon." The surrounding country is called Charo cha Mwesi, which means "Land of the Moon." At the same time the Kagera, which had been called by Stanley the Alexandra Nile, may certainly be counted as the southernmost and one of the principal sources of the Eastern Nile.

In England the theory of Dr. Baumann, in its general outline, has been accepted by Sir Clements Markham. Neither, indeed, has failed to recognize the objection that the small importance and low altitude of the Missóssi ya Mwesi scarcely justify so far-reaching a celebrity. The natives of the Unyamwesi are certainly unconscious of the existence of the "Mountains of the Moon" in their country. Years ago, in fact, Speke heard from them a tale of a marvellous mountain situated to the north of Kasagwe, a region to the west of the Victoria Nyanza. This mountain was said to be so high and so steep that no one could ever possibly ascend it, and to be rarely visible because it soared up into the clouds from which a pure white substance was wont to fall upon it.

The Discovery of Ruwenzori.

Other geographers, such as Hans Meyer and Ravenstein, attempted to prove that Ptolemy meant to indicate the mountains which form and encircle the Abyssinian tableland. In the "Proceedings of the Royal Geographical Society" for 1901 (p. 513), may be found an interesting discussion which followed the lecture of H. Schlichter on this subject.

The Italian geographer Prof. L. Hugues has contributed a note* on this question. He has attempted to bring the limited knowledge which was attainable in the days of Ptolemy as to the geographical position of both the lakes and the mountains into harmony with the more precise information at our command at the present day, by taking into account errors in the mensuration of longitude and latitude inevitable at the earlier period. The conclusions at which he arrives are entirely in favour of Stanley's view.

CARAVAN ON THE MARCH.

---

* *See* Appendix A.

Chapter I.

Stuhlmann, Scott Elliot, Moore, Johnston and, in fact, all the others who have visited and explored the range of Ruwenzori after Stanley, have accepted his interpretation of Ptolemy's text. Indeed, unless we admit that the ancient geographers must have had in one way or another some concealed source of information as to the facts, we find ourselves under the necessity of regarding them as gifted with prophetic powers. Be that as it may, the legend of the " Mountains of the Moon" is a tale of the past, and Ruwenzori, established at last in its own exact place on the map, known in all the details of its structure, measured in every dimension, no longer runs the risk of being lost from the memory of man.

We will now return to the story of its exploration.

On the 1st of June, 1888, after his first sight of the snow-clad summits of Ruwenzori, Henry Stanley was forced to return on his track, and, re-crossing the vast forest of the Congo, to join his rear-guard camp, where one of the ghastliest tragedies recorded in the history of African exploration had taken place in his absence.

He did not return to Ruwenzori until the next year, 1889, when he skirted the whole western slope of the range. He then traversed the plain between Lake Albert Edward and the mountains, and, turning northwards, followed their eastern slopes as far as the head of Lake Ruisamba. He thus spent more than three months, from April to July, in the immediate neighbourhood of the range, and saw the snowy peaks again and again. Wishing to gather more accurate knowledge of the shape and structure of these mountains, he dispatched Lieutenant W. G. Stairs on a journey of exploration among them.

Lieutenant Stairs left the camp of Bakokoro, 3,860 feet above

The Discovery of Ruwenzori.

the sea-level, about the beginning of June. He followed one of the north-western valleys of the chain for two days, steering for two rocky peaks of a striking conical shape (Twin Cones) which had been marked from a distance on the north-west extremity of the range. He reached an altitude of 10,677 feet, about 1,500 feet below the rocky peaks. Here he came in sight of a snowy peak which he estimated at 16,600 feet, but which was not the highest point.

Lieutenant Stairs had not an equipment sufficient to enable him to remain several days in the mountains. He was therefore obliged to cut his exploration short and descend to the plain. From the appearance and shape of these mountains he thought it possible to maintain their origin to be volcanic.

NATIVE PORTERS, UGANDA.

Chapter I.

In the month of December, 1899, Stanley at last finished his venturesome journey and reached Zanzibar with Emin Pasha and his followers. A few months later, Emin Pasha, at the head of a German expedition, set forth again to return to the equatorial lakes. About the beginning of June, 1891, he found himself once more on the western slope of Ruwenzori, encamped at Karévia, near the southern course of the Semliki (Issango) river.

It was from this encampment, 4,364 feet of altitude, that Dr. F. Stuhlmann, one of the members of the expedition, made a five days' excursion up the valley of Butagu, one of the largest of the western valleys of the chain. He reached an altitude of 13,326 feet, not very far from the snow, in sight of two snowy mountains.

He was obliged to return, owing to his limited means of transport and to the sufferings of the natives from cold. A good naturalist, a first-rate explorer and a painstaking observer, Stuhlmann was the first to give an accurate description of the successive zones of vegetation in its varying forms at different altitudes. He proved clearly that Ruwenzori is not a single mountain, but a real range. He distinguished four principal groups to which he gave, proceeding from north to south, the names Kraepelin, Moebius (the highest peak called Kanjangungwe by the natives), Semper (Ngemwimbi of the natives), and Weismann. He was able to photograph two of these groups from the upper Butagu Valley. He also showed that Stairs' suggestion of a volcanic origin for the range is without foundation. Strange as it may seem, he failed to recognize the presence of true glaciers, but was rather inclined to regard them as mere accumulations of snow.

Stuhlmann was succeeded in the exploration of Ruwenzori by the naturalist G. F. Scott Elliot in the years 1894–95. He

The Discovery of Ruwenzori.

made five expeditions by various routes towards the summits, through the Yeria, Wimi, Mobuku, and Nyamwamba Valleys on the eastern slope, and through the Butagu Valley on the western slope. He pushed his way up to the heads of the Yeria and Wimi Valleys, and reached his greatest altitude (13,000 feet) in the Butagu Valley.

Stricken with malarial fever, lacking means of transport, Scott Elliot gave proof of admirable tenacity, but was unable to gather any data regarding the region of the snows. His most interesting observations are with regard to traces of ancient glacier action in the valleys of Mobuku, Nyamwamba, and Butagu, which prove that they were at one period filled by vast glaciers. Like Stuhlmann, he excludes all possibility of a volcanic origin for these mountains. The most important results of Scott Elliot's exploration are botanical.

After Scott Elliot we have no further record of Ruwenzori for five years, although the period of exploration had been succeeded in Uganda by the period of European occupation. The whole time and the entire energies of the English military and civil services were required to deal with serious difficulties, and with the necessity of facing dangerous complications which seemed at times to menace the very existence of the newly established Protectorate. It was necessary to depose kings, and to put down revolts with such means, slender and insufficient at best, as were available from a coast several months' journey distant.

Thus we reach 1900 without any further addition to our knowledge of the range. In the spring of this year C. S. Moore, at the head of a scientific expedition for the purpose of studying the fauna of the great lakes, reached the eastern slopes of Ruwenzori with the intention of attempting the ascent.

11

Chapter I.

He had purposed to go up by the Nyamwamba Valley, which, however, he failed to reach owing to the impossibility of fording the River Mobuku, at that time swollen by heavy rains and quite impassable.

PORTERS' HUT, UGANDA.

As early as 1894 Captain (now General Sir Frederick) Lugard had pointed out to Scott Elliot the Mobuku Valley as the best route by which to reach the snow. C. S. Moore now started up by this route, taking with him a small number of Suahili porters besides a few natives of the valley. In spite of unfavourable weather, he was able to ascend the valley as far as its head, and discovered for the first time the glaciers which encircle its upper end. He went up one of these and reached the edge of the terminal ridge, 14,900 feet above sea-level.

C. S. Moore thus gave us the first irrefutable proof of the existence of genuine glaciers upon Ruwenzori. He further-

The Discovery of Ruwenzori.

more confirmed Stuhlmann's description of the range, recognizing from the east side the same distribution of the peaks into four principal groups.

Some three weeks later, Fergusson, who had left England with Moore but had been delayed by fever at Fort Gerry (now Fort Portal), proceeded up the Mobuku Valley and ascended the glacier to the height of 14,600 feet.

Shortly after Fergusson, Bagge, who was employed in the Civil Service of the Toro district and had already made an excursion up the valley of the Nyamwamba as far as the bamboo zone, pushed up the Mobuku Valley and reached the glacier. Bagge had a rough path cut by the natives up the valley, which proved a useful guide to subsequent explorers.

Sir Harry Johnston, High Commissioner of the Protectorate, accompanied by Messrs. Doggett and Vale, followed this track in September of the same year. His choice of this route was determined partly by the relative facility with which explorers since Moore seemed to have reached the glaciers, and partly by his conviction that the principal groups of the range were in the immediate neighbourhood of the head of the Mobuku Valley. Sir Harry Johnston's expedition reached the glacier and ascended to a height of 14,828 feet, but was unable to reach the ridge.

Johnston rebaptized the peaks under the names given them by the natives of those valleys, which were, of course, totally different from the names reported by Stuhlmann from the west of the chain. Thus, the Ngemwimbi, or Semper of Stuhlmann, became Kiyanja, while another peak visible from the Mobuku Valley received the name of Duwoni. Johnston succeeded in taking good photographs of the valley, of the Mobuku Glacier and of some peaks. He gave us also a detailed description of

13

Chapter I.

the flora and fauna of the mountain district. Like Scott Elliot, he noted traces of glacial action in the Mobuku Valley, about 3,000 feet below the point where is now the snout of the glacier. Like all his predecessors, he complains of an extraordinary persistence of bad weather.

A CAMP.

Thus in the year 1900 alone the Mobuku Valley had been explored by four separate parties.

In August of the following year, W. H. Wylde and Ward went up the valley and appear to have reached the ridge on the top of the glacier at the same altitude which was reached by Moore, about 15,000 feet. During the two following years we have no further reports of the Mobuku Valley until the expedition of the Rev. A. B. Fisher, who, with his wife, in January, 1903, arrived at the point which had been reached by Sir Harry Johnston.

The Discovery of Ruwenzori.

The geographical periodical "Globus," published, in 1904, a brief notice of an ascent on the chain of Ruwenzori made in April of that year by Dr. J. J. David. He was reported to have reached an altitude of about 16,400 feet. Two years later the "Bollettino" of the Italian Geographical Society published an article by Revelli on Dr. David's expedition based upon his own notes. David had chosen the route of the Butagu Valley to the west of the chain which had not been explored since Scott Elliot. Ascending a tributary valley he reached the glaciers in seven days. Through the glaciers he reached a col, covered with ice, upon which was a small rocky peak of gneiss about 150 feet high. From here he was able to see the valleys descending on the opposite slope towards Uganda. He was stopped here by the evident danger of proceeding alone upon the glacier. The altitude of the pass which he had reached (16,400 feet) seems to have been ascertained by triangulation. The absolute lack of more precise data and of any detailed account of the route which he followed prevents us from identifying the peak which he ascended. Probably David might have been able to point out his route upon one of the photographs taken by Stuhlmann in the higher Butagu Valley, and reproduced in his book.*

In the course of the same year, 1904, M. T. Dawe made an important botanical expedition up the Mobuku Valley. This expedition was, however, without results from the point of view of the mountaineer.

During the time which had now elapsed since the occupation, a railway line had been opened between Mombasa, on the coast

---

* There is the possibility that David may have reached the saddle between the Elena and Savoia Peaks of Mt. Stanley (about 15,750 feet), where a rocky tooth would seem to correspond to his description.

of the Indian Ocean, and Port Florence, on the eastern bank of the Victoria Nyanza. Port Florence was in its turn connected with Entebbe, the capital of Uganda, by a regular service of steamers. It had thus become possible to reach the centre of the African continent without difficulty, at small expense, and with an immense saving of time. The country was henceforward in a state of peace and security.

NATIVE CHIEF WITH HIS FAMILY, UGANDA.

After the expedition of David, no explorer availed himself of these favourable conditions until the end of the year 1905, when interest in Ruwenzori seemed suddenly to reawaken. Thus it happened that at the very moment when H.R.H. the Duke of the Abruzzi was forming his plan for an expedition in this region, and in the beginning of 1906, when he had actually

The Discovery of Ruwenzori.

taken such measures for carrying it into effect, as collecting material and preparing details of equipment, the range was already being attacked by determined mountain climbers bent upon rending the veil of mystery which had so long shrouded its secret.

In November, 1905, for the first time in the history of Ruwenzori, a party of expert mountaineers, Douglas W. Freshfield and A. L. Mumm with the guide Moritz Inderbinnen of Zermatt, arrived in the Mobuku Valley. They found the season especially unfavourable. After waiting for a long time at the upper end of the valley they were forced, by uninterrupted rains, to abandon the undertaking. They had succeeded in making one attempt, in the course of which Mumm had ascended the glacier, but without reaching the ridge.

In January, 1906, the Rev. A. B. Fisher, with his courageous wife, went up the Mobuku Glacier for the second time. In the same year an Austrian mountaineer, R. Grauer, with two English missionaries, H. E. Maddox and the Rev. H. W. Tegart, who during the preceding year had attained to an altitude of 14,000 feet on the Mobuku Glacier, climbed the high terminal ridge of the valley which had not been reached since 1901. They ascended to the summit of a small rocky peak which rises on a depression in the ridge to a height of 15,000 feet above the sea. This peak Grauer named after King Edward.

Finally, in October, 1905, a scientific expedition, sent out by the British Museum to study the fauna and flora of Ruwenzori, started from London under the direction of H. B. Woosnam. The other members of this expedition were G. Legge, R. E. Dent, M. Carruthers and A. F. R. Wollaston, a member of the Alpine Club. This expedition spent several weeks in the Mobuku Valley to collect scientific material, and

Chapter I.

in the interval made expeditions up the glaciers at the head of the valley.

First Woosnam alone, then a party consisting of Wollaston, Woosnam and Dent reached, in February, of 1906, the spur of rock on the ridge where Grauer, Maddox and Tegart had gone in January.

A few days later Wollaston and Woosnam attempted to scale the peak which rises above the western slopes of the valley (the Kiyanja of Johnston), but owing to the dense fog they were stopped on a knob of the ridge at an altitude of 16,125 feet, a little below the actual peak.

A CHIEF'S DAUGHTERS, OF BAHIMA STOCK.

The Discovery of Ruwenzori.

On the 1st of April, Wollaston, Woosnam and Carruthers, still pursued by bad weather, ascended the rocks beside the Mobuku Glacier and reached a rocky peak 15,893 feet high, which rises to the north-east of the valley, and which they believed to be the Duwoni of Johnston.

Two days later, the same party repeated the ascent of the rocky knob on the ridge of Kiyanja, and the reading of the boiling-point thermometer gave them this time a somewhat higher altitude (16,379 feet).

The persistent bad weather which hampered them on all these expeditions barely allowed them to perceive that other peaks of the chain rose up towards the north-west, and that they seemed higher than those which they had themselves ascended.

Before the departure of the Italian expedition, only vague and inaccurate reports of these ascents had come from Uganda. Nor had any precise and direct intelligence been received from the members of the British Museum Expedition.

To ensure a clear understanding of the facts, I have made out a table of all the explorations of Ruwenzori, which preceded the expedition of H.R.H. the Duke of the Abruzzi. In this table I have given the altitudes as reported by each writer. They are to be taken as approximate only, because none of them are drawn from a series of observations carried out with the precautions and the corrections necessary to an exact result. It is possible that, in addition to the expeditions which I have recorded, others may have been made by English residents in the protectorate. Of any such I am ignorant, as no account of them has been published.

Chapter I.

EXPLORATIONS OF THE RUWENZORI RANGE FROM ITS DISCOVERY, 1888, UP TO APRIL, 1906.

Date.	Explorers.	Route followed.	Point reached.	Height, feet.
June, 1889	W. G. Stairs	Russirubi (?) Valley	—	10,677
„ 1891	F. Stuhlmann	Butagu V.	—	13,326
„ 1895	G. F. Scott Elliot	Yeria V. Wimi V. Mobuku V. Nyamwamba V. Butagu V.	— — — — —	— — — — 13,000
„ „	Stephen Bagge	Nyamwamba V.	Bamboo Zone ...	—
Mar., 1900	C. S. Moore	Mobuku V.	Terminal crest...	14,900
April, „	Fergusson	„	Mobuku Glacier	14,600
May, „	S. Bagge ...	„	„	—
Sept., „	Sir H. Johnston, W. G. Goggett, and Wallis Vale	„	„	14,828
Aug., 1901	W. H. Wylde, and Ward	„	Terminal crest...	14,900
Jan., 1903	Rev. A. B. Fisher, and Mrs. Fisher	„	Mobuku Glacier	—
April, 1904	J. J. David	Butagu V.	Col on the top of the watershed (?)	16,400(?)
? „	W. T. Dawe	Mobuku V.	Base of the Mobuku Glacier	—
? 1905	H. E. Maddox, and Rev. H. W. Tegart	—	Mobuku Glacier	14,000
Nov., 1905	D. W. Freshfield, and A. L. Mumm	Mobuku V.	„	—

The Discovery of Ruwenzori.

Date.	Explorers.	Route followed.	Point reached.	Height, feet.
Jan., 1906	Rev. A. B. Fisher, and Mrs. Fisher	Mobuku Valley	Mobuku Glacier	—
„ 1906	R. Grauer, H. E. Maddox, and H. W. Tegart	„	Terminal ridge (King Edward's Rock)	15,000
Feb., 1906	H. B. Woosnam ...	„	„	„
„ „	A. F. R. Wollaston, H. B. Woosnam, and R. E. Dent	„	„	„
„ „	A. F. Wollaston, and H. B. Woosnam	„	Knob on the ridge of Kiyanja	16,125
April, 1906	The same, with M. Carruthers	„	Peak on N.E. side of the valley believed to be the Duwoni of Johnston	15,893
„ „	The same party ...	„	Knob on the ridge of Kiyanja	16,379

As may be seen by this table, the Mobuku Valley was the route chosen by the greatest number of the expeditions. By this route the high terminal ridge had been reached three times. By this route Wollaston and his companions had succeeded in ascending two peaks of the chain, one of which was estimated at over 16,000 feet high. It would appear that Dr. David must have reached a still higher point by the western slopes; but as I have already mentioned, the accounts of his expedition, which are at our disposal, are so vague as to render it impossible to identify the col which he ascended.

Chapter I.

NATIVE MARKET IN UGANDA.

Certainly no one of the forerunners of the Duke of the Abruzzi had succeeded in actually exploring the chain, that is to say, in obtaining the comprehensive knowledge requisite in order to describe the general formation of the line of watershed, the configuration of the mountains, the relative height of the different peaks, their connection with the several valleys, and the extent and position of the glaciers.

The chief explorers had observed that the snow-peaks fall into four distinct groups. It was not known, however, whether these groups were connected by high ridges or divided by deep valleys.

In the absence of systematic exploration it was impossible to bring into relation to one another the different descriptions of peaks as seen from diverse points. The confusion between the several names given to them increased the difficulty of comparing the different reports. In addition to all this, it had been given to only a very few, on rare occasions and from a great distance, actually to have sight of the whole chain.

INCENSE TREE (?)

Under these circumstances, the views of travellers as to the height of the principal peaks, the extension of the glaciers, and the general features of the range were widely divergent and based upon scanty foundations. These views were either derived from a fleeting vision of far-off mountains in great part hidden by lower buttresses of the chain, or else based upon knowledge of one single valley whose surrounding barrier of ridges hid from sight all the greater and more important features of the range.

Chapter I.

Thus it came about that the estimated altitude of the highest peaks varied between 15,000 and above 18,000 feet. The only trustworthy measurement was that derived from triangulation applied to the peak which appeared to be the highest. The triangulation was taken from various points to the south-east of this peak by Lieutenant Behrens of the Anglo-German Boundary Commission. The calculations based upon this triangulation gave a height of 16,757 feet. Colonel C. Delmé Radcliffe, however, who was at the head of this Commission, noted the possibility of higher peaks existing further northward and not visible from the point of observation.

The chain of Ruwenzori, without doubt the most important group of mountains and glaciers in Africa, and the one mystery still unexplored with regard to the question of the sources of the Nile, promised a fertile field for research. The arduous character of the undertaking and the uncertainty as to what obstacles might be encountered were calculated rather to attract than to dissuade so ardent an explorer and so keen a climber as H.R.H. the Duke of the Abruzzi.

CHAPTER II.

PREPARATION AND DEPARTURE OF THE EXPEDITION FROM ITALY. FROM NAPLES TO ENTEBBE.

Choice of Season—Objects of the Expedition—Organization of the Expedition—Departure from Naples—Mombasa—Lieutenant E. Winspeare falls ill—The Uganda Railway—The Tableland of Athi—Rift Valley—Port Florence—Kisumu Market—On the *Winifred*—The Gulf of Kavirondo—Sunset on Lake Victoria—Steaming upon the Equator—The Islands—Sleeping Sickness—The Archipelago of Sessé—Arrival at Entebbe.

 OF all the difficulties with which most of the predecessors of the Duke of the Abruzzi had had to contend in the exploration of Ruwenzori, the gravest had been the obstinate bad weather, the nearly incessant rain, and, in the brief intervals, the dense mist which shrouded the whole region. These untoward conditions gave the utmost importance to the decision as to which season seemed likely to be the least unfavourable. Judging indeed by the reports of former expeditions which had set forth in nearly every month of the

Chapter II.

year, it seemed that one single rainy season reigned supreme among the mountains without any break of fair weather.

MOMBASA—PORT KILINDINI.

The fact is that the great mountain range, rising like an island from the vast marshy plains of Uganda and the boundless forest of the Congo, becomes a centre of attraction for the mass of vapours sucked up by the tropical sun, which, condensing around the frozen peaks, form a permanent veil of fog and cloud. Thus it has come about that many a traveller has spent months and months in the immediate neighbourhood of the chain without once having sight of the peaks, or, at best, fugitive glimpses only.

Stairs and Stuhlmann in June, and David in April appear to have found climatic conditions slightly more tolerable than the other explorers. Wollaston, however, had very bad luck in April. Sir Henry Stanley writes in May that he saw the snow peaks uncovered " frequently, almost daily." On the other hand, Sir Harry Johnston cites the local residents to the effect that the best months are November and December. Freshfield, on the contrary, encountered the very worst weather in these months.

Accounts given by the missionaries residing in the district of Toro, at the foot of the chain, appeared to concur with Sir William Garstin's report on the basin of the Upper Nile in admitting the rains to be somewhat less frequent in January and February, and in June and July, although the atmosphere upon the mountains remained gloomy even during those months. The

MOMBASA—PORTUGUESE FORT.

experiences reported by pioneers were not in agreement with these forecasts which, however, it seemed better to accept, as forming the only basis for a decision. The Duke therefore

Chapter II.

arranged to leave Italy in the spring in order to reach the mountains by the beginning of June.

The expedition was accordingly equipped in the early months of 1906. Care was taken to enable it to reap the utmost profit from the exceptional opportunities for scientific investigation offered by a journey among mountains still almost unknown, and

THE UGANDA RAILWAY.

situated in peculiar conditions in the centre of a continent where much still remains to be discovered.

The main object of the expedition was of course geographical in the strict sense of the word; that is, to clear up the topography of the chain and ascertain the heights of the mountains. This work was to be completed by observations on physical geography, meteorology and magnetism.

In connection with these aims it was important to illustrate amply by photography the regions traversed. Next in order came geological and glaciological research, including mineralogy. Lastly, there was every reason to expect that in the yet unexplored valleys and mountain slopes interesting additions would be found to our knowledge of the fauna and flora of Ruwenzori.

With a view to carrying out this plan of research, the Duke of the Abruzzi selected as companions on this occasion Commander Umberto Cagni, who had taken part in both of his previous expeditions, and his aide-de-camp, Lieutenant Edoardo Winspeare, to assist in the topographical work and the observations connected with physical geography. The photography was entrusted to Cav. Uff. Vittorio Sella, who had already accompanied H.R.H. in the ascent of Mt. St. Elias. Dr. Achille Cavalli Molinelli, R.N., who had taken part in the Duke's Arctic Expedition of 1900, was once more chosen as medical attendant to the party, and was also to assist Dr. Alessandro Roccati in the collection of zoological and

SUGAR-CANE VENDORS.

botanical specimens. Dr. Roccati was furthermore specially entrusted with the geological and mineralogical researches.

Chapter II.

It would be a waste of words to discuss the necessity of taking Alpine guides on an expedition of which climbing was to form the essential feature. Furthermore, as there was every reason to expect that, in the course of the

AT A RAILWAY STATION.

exploration of a mountain range whose summits had been estimated by previous travellers at heights varying from 16,000 to over 20,000 feet, prolonged sojourns above the snow-limit would be necessary, the expedition had to be furnished with the needful equipment for glacier camps, more or less on the lines followed in the ascent of Mt. St. Elias in Alaska. This involved the necessity of taking out European porters as well as guides, for it was impossible to count upon the services of the natives beyond the foot of the glaciers.

The guides chosen for the expedition were Joseph Petigax, the intelligent and devoted companion of the Duke in the Alps, in Alaska, and on the Polar Expedition, and César Ollier. Both of these guides, as well as the porters, Joseph Brocherel and Laurent Petigax, were from Courmayeur, in the Valley of Aosta. Ollier and Brocherel had already guided the Mackinder-Hausburg Expedition to Mount Kenya (East Africa) in 1899. There was also Erminio Botta, of Biella, the assistant photographer of Sella, who had had experience of rough life and exploration in the Caucasus, in Alaska and in the Himalaya. Lastly there was the cook, Igino Igini, of Acquapendente, who had passed an Arctic winter in the Duke's service in the Bay of Teplitz.

I will not describe at length the preparation of the equipment, to which the Prince attended with the same scrupulous care for detail which had so much contributed to the success of his former undertakings. It was especially difficult and complicated on this occasion, owing to the necessity of meeting the requirements both of a long march

PORT FLORENCE.

across tropical regions, and of a prolonged sojourn among ice and snow.

Everything was ready by the beginning of April, and on the

Chapter II.

evening of the 16th the whole party set out from Naples on board the German liner *Bürgermeister* bound for East Africa.

THE "WINIFRED" ALONGSIDE THE PIER IN PORT FLORENCE.

The distance from Naples to Mombasa, situated on the East Coast a little over four degrees south of the Equator, is about 4,100 miles. The steamers usually take seventeen days on the voyage, with brief stops at Port Said, Suez, Aden, and Jibuti.

Africa as seen from the Red Sea is far from attractive. The coast is low and sandy, flat or bounded by dunes. The hills are barren and naked, the country baked by the sun, desolate and sterile. The ports upon the high road of the great trade lines, present a profoundly depressing spectacle. Arabs, Turks, and negroes in rags and squalor, with swarms of crippled, diseased, and leprous beggars combine to form a population of countless races, poisoned and deteriorated physically and morally by sudden contact with a civilization too widely different from their own. The white man's highly complicated and subtle civil organization, the growth of an immeasurably long period, during which individual development has kept pace with the evolution of the body politic, has been suddenly thrust with disastrous results upon races undeveloped and unprepared for its reception.

The voyage was most prosperous, with calm sea and fine weather which changed only toward the end. Professor Koch was among the passengers, and the monotony of the voyage was thus pleasantly broken for the Duke and his companions by talk about Equatorial Africa, whither Koch was returning to proceed with his studies on the sleeping sickness, that ghastly scourge which has in a few years nearly depopulated vast districts around the great lakes.

The last days of the journey were clouded by the illness of Lieutenant Winspeare. High fever developed and it soon became manifest that it would be impossible for him to

NATIVES GOING TO MARKET, KISUMU.

accompany the expedition across the unhealthy region between the coast and the mountains, which is often fatal even to those who undertake the journey in a perfect condition of health.

33

Chapter II.

Mombasa is situated upon an island surrounded by a steep coast of coral formation, and covered with palms. This island lies in a bay forming two sheltered channels, of which the

A SHED IN THE MARKET PLACE, KISUMU.

easternmost, known as Port Mombasa, is narrow and difficult of navigation, and suited only to the small craft which trade along the coast, while the other, known as Port Kilindini, is capacious and convenient, and here the English, French, German, and Austrian liners ride at anchor. The *Bürgermeister* entered this port on the morning of 3rd May.

Concessions made by the British authorities with regard to customs and transport, and their courteous assistance, together with that of the few Italian residents, facilitated the landing of the equipment.

Mombasa, like the other cities of this coast, was founded about the year 1000 by Arabs and Persians. Coins, inscriptions, and architectural fragments record their occupation.

From Naples to Entebbe.

The island upon which the city is built is known in the native language as Kisiwa mwita, or " Island of War," a name which agrees with its history, which is entirely °made up of warlike vicissitudes. Mombasa is the best port on the whole east coast of Africa, and was a valuable station on the old trade route for India before the Suez Canal. For these reasons it was for centuries one of the most eagerly sought positions and one of those most persistently disputed between the Arabs, Portuguese, and Turks, who held it alternately. When the Portuguese domination came to an end in 1729, Mombasa was

KAVIRONDO WOMEN.

governed for over a century by Arabs of the family of Mazrui, under the nominal suzerainty of the Imans of Oman. When the latter transported their capital from Maskat to Zanzibar,

Chapter II.

they drove out the Mazrui from Mombasa and re-established their own effective domination there in 1837.

The Portuguese domination is recorded by the ancient fort, a massive edifice built towards the end of the sixteenth century, several times dismantled, but which still stands and bears cut in the stone the Christian symbol " I.H.S.," together with the eagles of the Austro-Spanish dynasty which governed Portugal in 1635, when the fort was restored. It now contains the prisons and a military store-house.

After 1848, English and German geographical exploring expeditions followed one another. In their wake were formed colonial trading companies, which established themselves on the coast and penetrated the country little by little, gradually obtaining concessions from the Sultanate of Zanzibar or through treaty with native chiefs. Anglican and Catholic missionaries next made their way still further into the interior, where they had been preceded by the Mohammedans. In consequence, religious wars lacerated Uganda for many years. In 1890, Germany and England established by treaty their respective zones of influence. Three years later, the Imperial Government of Great Britain took over the protectorate, and since then has pushed the occupation up to the boundaries of the Congo State.

On the morning of the 4th of May, Lieutenant Winspeare was carried to the hospital, which stands on a healthy, airy, and pleasant site overlooking the ocean and the picturesque Portuguese fort. The grey sky and fine rain seemed to fit the depression which all felt at having to leave a comrade behind at the very outset of the undertaking. Lieutenant Winspeare recovered sufficiently to leave Mombasa to return to Europe on the 12th of May.

From Naples to Entebbe.

As is well known, Mombasa is now connected with Lake Victoria by a railway which runs north-west from the coast in a direction nearly parallel to the Anglo-German boundary, and

KISUMU MARKET.

touches the lake at Port Florence at the head of the Bay of Kavirondo, almost upon the equator.

At eleven o'clock on the morning of the 4th of May, the Italian expedition left Mombasa by the railway, which traverses regions completely unknown less than thirty years ago.

The distance from the east coast to Lake Victoria is 584 miles. At the present day this distance is covered in a couple of days without the least fatigue, comfortably seated in the little narrow-gauge railway carriages which are arranged inside like those on our own Sardinian railways. Only a few years ago several months of difficult and dangerous travel among warlike tribes, over wretched tracks, in an unfavourable climate and with all the complications, obstacles and expense

37

Chapter II.

of a numerous caravan of porters were required to cover this ground.

The construction of this railway was a truly great work, owing to the serious obstacles which had to be overcome. It stands as a witness to splendid perseverance and resolution in an incessant struggle for six years against the gravest difficulties. Vast tracts of the country are absolutely lacking in water or resources, and, in fact, practically a desert. A great part of the way passes over mountains where the line rises to a height of 7,700 feet, descends to 6,000, and mounts again to 8,300, only to drop down to 3,700 on the shore of the lake.

KISUMU MARKET.

No help was forthcoming from the natives, mere naked savages, devoid of industry or skill, incapable of work in any shape whatsoever. It was necessary to transport an army of

From Naples to Entebbe.

20,000 labourers and artisans from India, to feed, lodge and clothe them, and to supply them with the necessary implements. Everything had to be brought out either from England or from India, thus necessitating as great forethought and as complete organization as are required for a military campaign. In addition to all this, owing to the enormous difficulty of transport in a country where the tsetse-fly makes the use of beasts of burden impossible, the work of cutting and preparing the line could only proceed a very short distance in advance of that of laying the rails. Some of the districts traversed were unhealthy. At times epidemics prevailed. The men were tormented by divers kinds of parasitic insects. Lions made numerous victims and struck terror into the workmen.

The undertaking was commenced in 1895, before the completion of the conquest of Uganda. In the very midst

WITHIN THE ENCLOSURE OF THE MARKET, KISUMU.

of the work in 1897, the colony ran a serious risk through mutiny among the Sudanese troops and the rebellion of the Kings of Uganda and Unyoro, instigated by the Mohammedan

Chapter II.

party. Yet in the third year after its inauguration, the Uganda Railway counted 179,000 passengers.

A European, landed for the first time in Africa, must experience a strange sensation on finding himself suddenly transported by railway into the very midst of a landscape, where every feature, people, animals and plants unite to form the picture which he had so often attempted to create by imagination.

Immediately after crossing the bridge that joins Mombasa to the continent, the railway begins its ascent to the tableland, passing first through fields of mango, cocoanut, banana and all the beautiful vegetation of the coast zone ; next, through the undulating and bare plains of the Taru desert, where thorny bushes and a few euphorbias are the only plants ; then once more through a fertile country among flowering fields and woodlands.

The stations, placed at intervals of 20 miles from one another, consist each of a little wooden hut, beside a shed standing alone in the wilderness. Every 100 miles is a central station. Here the natives collect in numbers from the neighbouring villages to sell sugar-cane and bananas to the third-class passengers.

The train continues to climb by a gentle grade, and the snowy peaks of Kilimandjaro become visible to the south. The landscape is monotonous, and the country infested by the tsetse-fly. A little further on, for reasons unknown to us, the dangerous insect disappears, and a veritable Eden opens to the view of the traveller. This is the Tableland of Athi, the famous game preserve of the Government, upon whose rich pastures, dotted with umbrella acacias, graze peaceably, almost without fear of the train, whole herds of zebra, buffalo, gnu, antelope, and gazelle. Giraffes, too, may be seen peeping timidly

From Naples to Entebbe.

from behind the groups of trees, or ostriches driven into swift flight by the noise of the passing train; while now and again the traveller may be so lucky as to behold a lion sauntering

BANANA SELLERS, KISUMU.

across the plain, less startled perhaps than the onlookers, who gaze astounded upon the extraordinary sight.

About half-way on the journey stands Nairobi, a flourishing little town, thanks to the healthy climate and the fertility of the soil. Presently the country assumes a mountainous character, and the line climbs steep ridges clad with luxuriant forests of juniper and other evergreens, or penetrates into narrow silent valleys to reach at last the summit of the heights which form the eastern cliff of the Rift Valley, that vast entrenchment which winds through high plateaux between Lakes Rukwa and Nyassa to the south-west, and the Gulfs of Tajurra and of Aden

Chapter II.

to the north-east. The railway descends nearly 2,000 feet to reach the bottom of this valley, which is about 30 miles wide, and is dotted with tiny volcanoes, some active, others extinct. Lakes and ponds of sweet or salt water swarm with every species of aquatic bird, and the abundant water-courses make this one of the most fertile regions in Africa, as well as one of the most famous hunting grounds.

After Nairobi the line crosses districts inhabited by the Wa-Kikuyu, agricultural and sedentary tribes; and by the Masai, nomads and herdsmen, great breeders of cattle and bold warriors, who stopped the progress of many an explorer of old. The Masai as a race are finely proportioned, with a proud, fierce

OFF RUSINGA ISLAND.

mien and rather regular features, except the ears, which are unrecognizable, so deformed are they by absurd and voluminous ornaments. The women wear also heavy long copper spirals

42

wound around their neck, arms and legs. Their clothing consists of a mantle of stuff or of skins stitched together, fixed upon one shoulder after the fashion of a toga, or around the chest under the arms.

NATIVE CANOE WITH THE PROW OF PEACE.

After passing through a portion of this valley, and close to several little lakes set in an enchanting landscape, the railway proceeds to ascend the opposite forest-clad slope to a height of 8,300 feet, whence it again descends from valley to valley, through groves of acacias, bananas and palms, to the level grassy shores of Lake Victoria. The train makes straight for the pier of Port Florence.

While the goods were being carried on to the steamer there was time to pay a short visit to the market of Kisumu. Here the natives assemble in great numbers from the neighbouring

Chapter II.

villages, mere groups of huts surrounded by a hedge. They belong to the tribe of Kavirondo, which was formerly one of the most powerful and one of the wealthiest tribes around the Victoria Nyanza. The crowds of men and women come across the level country, carrying on their heads baskets woven with great art out of grasses. The young people of both classes go completely naked until marriage ; after marriage they wear a scrap of goat-skin over the hips, rather as a symbol of the conjugal state than as a garment. They are renowned for their modesty and for their morality, which contrasts with the dissolute tribes in the neighbourhood, although the latter are clothed. The Kavirondo are sober, gentle, peaceable and sociable. Sir Harry Johnston regards them as the most moral people of Central Africa.

The native costume is unfortunately doomed to rapid disappearance. Here, as everywhere else, civilization, intolerant of all forms, aspects or traditions of life that differ from its own, is swiftly introducing that monotonous uniformity which tends to turn the whole world into one people. It can scarcely be hoped that Kisumu, situated as it is at the terminus of a railway, will long preserve its distinctive character.

Clear indications of a rapid change are already visible. Mingling with the naked natives are many partially or even wholly clad in garments of white, striped, or gaily printed cottons, over which they often wear some hideous European garment, such as a waistcoat, a jacket, or a tail coat, without the least consciousness of their grotesque and absurd appearance.

The market is held in the open air or under sheds erected on purpose. It consists mainly of small traffic in dried fish, sweet potatoes, grain and bananas. The buyers stand in groups around the sellers, who crouch or sit on the ground beside the baskets of every conceivable shape which contain their wares. Men and women smoke the short straight pipes of the country. Others circulate hither and thither with that buoyant and elastic tread, like the gait of a wild animal, which comes from the habit of moving without the impediment of clothes. The women wear a string of beads around their waist, from which a sort of tail of woven fibres hangs down behind. The men wear necklaces of glass beads, with bracelets of iron on their wrists and their ankles. The mode of dressing the hair is frequently fantastic and embellished by feathers, hippopotamus teeth, etc., etc.

AMONG THE SESSÉ ISLANDS.

The current coin, as throughout Eastern Africa, is the rupee, worth about 1s. 4d. The use of cowries for currency persists only in those forms of trade which require subdivision to an infinitesimally low value.

45

Chapter II.

The steamship *Winifred*, with its twin, the *Sibyl*, performs a regular service between the harbours of Lake Victoria. The trade increased so rapidly that a third steamship was launched in 1907, and a fourth is already in construction.

Port Florence is situated in the little bay of Ugowe, at the eastern extremity of the Gulf of Kavirondo, opposite to and a little higher than Kisumu, which is on the other side of the bay.

The Gulf of Kavirondo runs inland to a distance of about 45 miles, while in some places its width scarcely reaches 3 miles.

NAPOLEON BAY, LAKE VICTORIA.

Its outlet into the lake is narrow and almost closed by islands The water of the gulf is yellow, dirty, and stagnant, nor is the least trace of any current perceptible. It is dotted with

46

floating islands formed by tangled masses of aquatic plants, upon which germinate and grow the papyrus, nymphæa and other species, which afford shelter to myriads of aquatic birds.

The north coast of the gulf forms a level plain. On the south side, at a short distance from the shore, a series of volcanic peaks, more or less rounded on the top, rise gradually into a chain of wild mountains, culminating in a jagged ridge, overtopped by a high and fissured cone nearly 4,000 feet above the lake.

Navigation on the Victoria Nyanza ceases with nightfall. On the evening of the 6th of May accordingly the *Winifred*, which had started at half-past two in the afternoon, cast anchor near to the Island of Rusinga where she was to pass the night, at the point where the Gulf of Kavirondo opens into the lake. The steamer was immediately surrounded by native canoes. These are large boats of slender form, carrying twenty rowers or more, not roughly hollowed out of tree-trunks, but built regularly from the keel upward with boards held together by fibre cords and the interstices caulked with fibre and resinous gums. The prow is armed with a long sharp point, covered at ordinary times by the " prow of peace," the extremity of which is turned vertically upward and is frequently adorned with feathers, horns of animals, etc.

The evening was now closing in. The slender canoes were leaving the sides of the *Winifred*. The fine nude torsos of the native oarsmen strained every muscle to the rhythmical stroke. The sharp click of the oars on the rowlocks was already dying out in the distance. The waves broke with a gentle murmur on the shore of the island. In the shallow water the hippopotami lifted their ungainly heads from among the reeds, while flights of birds sought their roosting-places on the scattered rocks with

Chapter II.

shrill cries. The sun was setting in a halo of fiery clouds. The last rays lit up the unfamiliar scenery where the shadows of night increased the sense of surrounding mystery.

SHORES OF LAKE VICTORIA, NEAR KAMPALA.

Little by little the colour of the sky passed from red and purple to colder hues and through subtler tones. Soon clouds and water and islands seemed to mingle and vanish in the twilight which was spreading swiftly over the surface of the lake.

For hundreds and hundreds of miles on every side stretched the vast regions of Central Africa, unknown up till yesterday, inhabited by that unhappy race which has survived a martyrdom of centuries, crushed under its fearful past of slavery, blood-thirsty rulers and murderous wars. To-day this era of violence has come to an end or is on the point of ending. The European, who for years past had bought ivory and slaves from the infamous Arab merchant, is now endeavouring to atone for the past and hopes to bestow a future of peace and prosperity upon the black by means of Christianity and civil organization. England has led the way in this heavy and laborious task, just as she had already taken upon herself the duty of routing out slavery, in the struggle she has carried on by sea and

BOTANICAL GARDENS, ENTEBBE.

land practically single-handed for eighty years. The goal, however, is still very far distant. Vast regions are wholly unexplored and out of all European control. Elsewhere the

Chapter II.

inferiority and weakness of the negro are too strong a temptation to his economic exploitation. In many places a state of social security appears to have bred sloth and dissoluteness among the natives, together with intemperance and lack of self-control, while every species of disease devastates the wretched and degenerate population. The civilized nations have but their own love of justice from which to draw the strength and consistency of purpose needful to carry on without hesitation a humanitarian work which demands disinterestedness and self-denial, gifts unfortunately rare in social aggregates.

On the morning of the 7th of May, by daybreak, the *Winifred* proceeded on her way, no longer in the muddy and colourless Gulf of Kavirondo, but in the open lake, whose waters are limpid and transparent, of a rich colour between emerald and blue, and as pure as crystal. A few hours after leaving the shore the land fades out of sight, giving the illusion of being on the high sea. The Victoria Nyanza is, in fact, surpassed in size only by Lake Superior in North America, and is so vast that it is possible to voyage along or across it for more than 200 miles without seeing land. It is like the sea, too, in its sudden and dangerous storms which raise up waves as high as those of the ocean.

The hydrographic survey of the shore was only finished last year, 1907. The shore line measures 3,200 miles and the survey occupied Captain B. Whitehouse seven years. The centre of the lake is still in great part unexplored and gives rise to numerous legends which are current in the country about islands inhabited by cannibals, ships swallowed by whirlpools, monsters which inhabit unexplored abysses, and other such matters.

50

From Naples to Entebbe.

Even when out of sight of land the voyage is never monotonous. The aspect of the sky varies unceasingly. Vapours and clouds perpetually form, and dissolve, or gather into dark

GOVERNOR'S HOUSE, ENTEBBE.

storms, while the water, reflecting their changes in endless variety of colour and tone, presents a spectacle which is never wearisome. Flights of swallows pass through the air. Swarms of minute gnats dance on the surface of the water like a light mist. These are the only tokens that land is not really far off.

The course of the steamers follows the equator, roughly speaking, and crosses the northern extremity of the lake from east to west, steering clear of the chain of islands which lie along the coast. These islands form a breakwater, sheltering a wide and practically continuous channel where canoes and small sailing craft can navigate in safety. They vary in size from mere rocks just rising above the water, and whitened by the deposit of aquatic birds, to islands so vast as to form a small region in themselves, clad with dense forests, girt about with irregular and deeply indented coasts, crowned with mountains attaining to a height of 2,000 feet above the surface of the lake, and inhabited by tribes which seem almost to have acquired special characteristics in their long separation from the mainland.

The most important and the greatest of all is the Island of Buvuma, one of the group which crowds around the entrance to Gulf Napoleon, and masks the exit of the Nile from the Victoria Nyanza. Buvuma was formerly inhabited by a warrior tribe which, relying upon a strong fleet, defended its independence with great valour against the powerful kings of Uganda.

The sleeping sickness has turned into vast graveyards the greater number of the beautiful and fertile islands of the

MARKET, ENTEBBE.

archipelago. After depopulating whole districts of the Congo, it appeared in Uganda between 1900 and 1902 and has spread further and further, following the main routes of communication, invading step by step the territories of the Baganda, Basoga and

Kavirondo, and making gigantic inroads even to the point of 40,000 victims in one year. The sleeping sickness is especially fatal to men in the prime of life, and hence whole villages and

MARKET, ENTEBBE.

islands may be found tenanted by women and children who alone have survived.

The British Museum Expedition mentioned in the preceding chapter, found in the district of Maniema, south of Ruwenzori, a multitude of natives stricken with the sickness and driven out from their villages, only to wander hither and thither in the country and die, untended, by thousands.

Hospitals have been started and are increasing in number. Attempts are made to organize help and to encourage emigration from the infected districts towards those which are still healthy, but the means are utterly inadequate to grapple with the swiftness and the activity of this fearful plague.

Sleeping sickness is caused by a parasite, a trypanosoma, discovered by Dr. Aldo Castellani in the brain of patients who

Chapter II.

have died of the infection. This trypanosoma is introduced into the system through the sting of a tsetse-fly, the *Glossina palpalis* (Col. Sir D. Bruce).

Occasionally the disease breaks out in the form of acute mania ; at other times its development is slow and insidious, with only a slight change in the appearance of the patient ; presently vertigo makes its appearance, with headache and swelling of the lymphatic glands of the neck. At last come trembling, somnolence, a quick pulse and an apathy which increases until it reaches the point of torpor or coma. No really sure remedy is known. Sundry arsenical preparations appear to be efficacious, one of them, *atoxil*, has recently given results which are rather more encouraging, but it is still uncertain as to whether a real cure can be expected.

IN THE COURTYARD OF THE EQUATORIAL HOTEL, ENTEBBE.

From Naples to Entebbe.

On approaching the north-west corner of the lake, the steamer enters a channel between the Sessé Islands and the coast. The Archipelago of Sessé, where concessions of land have

NATIVE HUTS AND PLANTAINS.

been granted to Italian companies for the cultivation of coffee and the collection of rubber, is the jewel of the Victoria Nyanza. Luxuriant forests cover the great islands down to the very brink of the lake, where the foliage is mirrored in the limpid waters. You would esteem it an earthly paradise, yet that charming scene conceals unspeakable desolation. The last forlorn remnants of the inhabitants, decimated by the dire disease, live mourning for their daily bereavements and dreading their impending fate.

The *Winifred*, soon after passing the entrance to Murchison Bay, entered the Bay of Entebbe and came alongside the pier about 3.30 p.m., May 7th.

The expedition had now reached the end of civilized means of communication, after travelling 4,750 miles in twenty-one days.

55

Chapter II.

The Duke of the Abruzzi was received on landing by the High Commissioner of the Protectorate of Uganda, Mr. Hesketh Bell,* who offered him hospitality in his own house, together with Commander Cagni. The other members of the expedition were entertained by Messrs. G. F. M. Ennis and W. M. Carter, both judges of the High Court of Uganda, and by Major L. C. E. Wyndham. The guides were put up at the Equatorial Hotel, kept by an Italian, Signore Berti.

---

* In October, 1907 the High Commissioner of the Protectorate received the title of "Governor."

CHAPTER III.

From Entebbe to Fort Portal.

Entebbe the Capital of the Protectorate—The Six Hills of Kampala—
H.H. Dandi Chwa, Kabaka of Uganda—The Missions—Commander Cagni's
Illness — Equipment — Formation of Caravan — Departure from Entebbe—
General Characteristics of the Country Traversed — Baganda Villages—
Climate—Baganda and Suahili Porters—Encampments—Visits of Chieftains—
Exchange of Presents—The Camp of the Blacks—Mitiana—Lake Isolt—The
Uganda-Toro Frontier—First sight of Ruwenzori—Butiti—King Kasagama—
Arrival at Fort Portal.

ENTEBBE, or Port Alice, founded
by Sir Gerald Portal barely
fifteen years ago, is the political
and administrative capital of the
Uganda Protectorate. The Protectorate includes much more than
the old kingdom of Uganda, having
been enlarged by the addition of
the kingdoms of Toro, Unyoro and
Ankole, which form a semi-circle
to the west and south of Uganda
proper. To these we must add the
district to the east of Lake Kioga and around the great
extinct volcano Elgon, as well as the vast regions known as
the " Nile Province " and the " Rudolf Province " to the north.

57

Chapter III.

The town is situated upon two hills at the extremity of a peninsula formed by two arms of the lake. The streets are wide and lined with houses, built according to the usage of the tropics, with wide verandas surrounded by gardens full of flowers. The site is enchanting, overlooking the great lake, dotted with picturesque islands; the wire netting, however, over windows, verandas and doors, tells its own story of malaria.

GENERAL VIEW OF ENTEBBE.

There is an hotel, there are Protestant and Catholic Churches, there are three hospitals, several commercial firms (among others a branch of the "Italian Colonial Society" established in Zanzibar), and sundry shops and stores kept by Indians and Goanese.

Along the shore of the lake stretches a considerable botanical garden, which contains a collection of the flora of the region as well as experimental cultivation of exotic plants, such as European vegetables, flowers and fruits, coffee, tea, cotton and even vines. Many of these are already cultivated widely and with good results in the Protectorate.

At the northern end of the town is the public market, the habitual haunt of the natives who congregate in great numbers around Entebbe hoping to get a job or an engagement as caravan porters.

SHED IN THE MARKET, ENTEBBE.

The native village, numbering some hundred huts in all, lies further inland, among plantations of bananas, fields of maize and lofty trees.

A fine and well-kept road leads from Entebbe to Mengo or Kampala, the native capital of the kingdom of Uganda, about 20 miles off, built upon a group of hills, each one of which is occupied by a different community. Mengo is the royal hill, Nakasero is the name of the hill where the English officials reside; the buildings and churches of the three different missions, one Anglican and two Roman Catholic, French and English, governed each by its bishop, occupy the three separate hills, Namirembe, Rubaga and Nsambya. Last

NATIVE HUTS, ENTEBBE.

of all comes Kampala, "the hillock which was contemptuously given to Captain Lugard by Mwanga, where the first seed was planted from which the British Administration all over these vast territories grew and prospered."*

* Sir Harry Johnston, "The Uganda Protectorate," Vol. I., London, 1904.

From Entebbe to Fort Portal.

The common centre of these diverse congregations is the bazaar, with shops well stocked with all sorts of wares, kept by Indians.

The present King of Uganda is His Highness the Kabaka Dandi Chwa, aged barely thirteen years. He was placed upon the throne in 1897, when his father Mwanga was deposed. His ascent to the throne was attended by the ceremonies consecrated by national tradition, with one important exception. That portion of the solemn and ancient ceremonial which consisted of a large slaughter of subjects, was on this occasion omitted.

THE KAMPALA ROAD, ENTEBBE.

The constitution is unchanged. The child King has three Regents by his side, the Katekiro, or Prime Minister, the Supreme Judge, and the Treasurer. He governs with the assistance of a Council composed of twenty chiefs of districts and of sixty-six notables who represent all the districts. The members of this Council are chosen by the King, but the representative of the British Government has the right of veto.

It is to be hoped that Dandi Chwa, carefully educated according to civilized principles of modern justice, may retain no trace of the bestial ferocity of his ancestors, and that the royal palace of Mengo may never again see such horrors as steeped it in blood in the days of the Kings Mtesa and Mwanga.

HANGING NESTS ON THE LEAVES OF A PALM.

Hundreds of human victims sacrificed at a word from a sorcerer, wholesale slaughter of the population for a whim, or on account of a dream, or to quiet the superstitious terrors of the Kabaka, torture, mutilation, daily murders of wives, of servants, of slaves, the country emptied of women to fill the harems of the kings or chieftains, all this formed a condition of

A ROAD IN UGANDA.

affairs whose incidents were so especially ghastly that they would seem to surpass the limits of human possibility if they were not proved by the unanimity of the descriptions of witnesses who saw Uganda in those days. The neighbouring kingdoms were in a similar condition, while the population of the islands were cannibals.

The transformation of the country in so few years is miraculous, and the greater portion of the merit is to be attributed to the Missions. These Missions are the direct continuation of the first Anglican Mission which came to Uganda in 1877 on the invitation of King Mtesa, transmitted to England by a letter of Stanley, which has become historical.

Chapter III.

This was followed two years later by the French Roman Catholic Mission. The persecution under Mwanga, the murder of Bishop Hannington, the torture and burning alive of many native Christians failed to put a stop to the work which progressed with extraordinary rapidity, undisturbed by the civil wars and political changes. In 1895, an English Roman Catholic Mission was added to the list.

The number of native converts to Christianity increased yearly by thousands, while Islamism remained stationary. Manners and customs rapidly improved. Education followed

NATIVE HUT.

moral training. The missionaries created a written language for the country where none had heretofore existed. Schools grew up by hundreds beside the churches.

BANANA PLANTATION AND HUT NEAR FORT PORTAL

From Entebbe to Fort Portal.

At the present time many villages around Entebbe and Kampala are entirely Christian. The blacks may be seen any day squatting on the ground around the catechist. Throughout the country numbers of natives may be met going or coming from their labour in the fields, praying or reciting the rosary on their way. They are all clad in the long white tunic with wide sleeves, which has almost universally replaced the older garment made out of strips of the bark of a special variety of fig-tree, beaten until they become soft and flexible, and stitched together with great art. On Sunday, in the spacious churches of Mengo, which afford room for several thousands of persons, men, women and childen may be seen worshipping with exemplary fervour and decorum.

On the other hand, it cannot be denied that Islamism exercised an important and beneficial influence in rescuing the country from its barbarous condition. In many districts the Mohammedans are still in the majority.

The Italian expedition remained at Entebbe from the 7th to the 15th May to prepare in detail the organization of the caravan. During this time Commander Cagni unfortunately fell a victim to the unhealthy climate, taking the malarial fever on the 8th of May. This persisted, in spite of quinine injections, and was complicated by intestinal inflammation. It soon became necessary to remove him to the hospital, which was situated in a healthier position, and afforded better accommodation.

Owing to this calamity, the Duke lost invaluable assistance at the very moment when the work began to be difficult and complicated.

The luggage of the expedition had been carried by porters to the courtyard of the Equatorial Hotel, followed by a crowd

65

of inquisitive children and adults. Here the cases were opened, and their contents verified and inventoried. The whole camp outfit, including tents, beds, sleeping bags, stools, tables, baths, cooking utensils, the hermetically sealed cases containing clothing ; the photographic materials, and the materials for the zoological, botanical and mineralogical collections ; the arms and ammunition, formed 114 loads weighing about 47 lbs. each, all numbered and so marked as to be immediately recognizable.

A HILLY BIT OF ROAD.

The commissariat formed 80 additional loads of the same weight, each one of which contained rations for 12 persons during one day. The supplies had been laid in on a calculation of a sojourn of 40 days above the snow-limit, and of a period of the same length below, to allow for the journey from Entebbe to the mountains and back. The rations were in tin boxes, soldered and enclosed in thin wooden boards. The only difference between the high-mountain rations and those for the lower regions was that the latter were without tinned meat, because it would be easy to find fresh meat supplies throughout the inhabited regions.

According to calculation 194 porters were needed to carry the entire equipment. In addition to these there were the caravan leaders, the personal servants, or "boys," with their own porters, the natives who were needed to take care of the horses and mules, and who were to drive the oxen, goats and sheep which were provided for the sustenance of the caravan,

ACROSS THE MARSHES.

and other natives, with sundry minor attributions. The total mounted up to above 300 persons.

Mr. J. Martin, Collector, who had special experience in organizing caravans and journeys, had caused the men to be selected and got together during the months preceding the arrival of the Italian expedition by Sig. Bulli, an ex-employé of the Italian Colonial Society, who was also to accompany the expedition.

Chapter III.

Three horses and three mules had been provided for occasional riding, beside two rickshaws holding one or two persons each, to be drawn or pushed by natives, for use on the relatively level portions of the road.

PAPYRI AND WATER LILIES.

Everything was ready by the 12th of May. The Duke, however, lingered three days more, as he could hardly make up his mind to leave Cagni behind. At last it became obviously necessary to set forth without him. The probable duration of his illness was too uncertain, and further delay would have involved the risk of letting the best season pass, not to mention the risk of some one else falling ill, and so endangering the whole success of the expedition. They could only hope that Cagni,

THE TROPICAL FOREST

From Entebbe to Fort Portal.

thanks to the devoted care of the excellent Doctor Hodges, might recover in time to overtake them. With this object in view, he was left provided with all the equipment necessary to permit of his setting out as soon as he should be sufficiently recovered.

On the 14th of May, H.R.H. and the rest of the party took leave of the kind hosts who had done so much to make their stay at Entebbe pleasant for them. The Collector, Mr. Martin, as representative of the Protectorate Administration, accompanied them as far as the frontier kingdom of Uganda with an escort of twenty-seven native soldiers and sixty-seven porters.

ELEPHANT GRASS.

Early on the morning of the 15th the porters with their caravan leaders, the boys, and the soldiers were gathered in the courtyard of Berti's Hotel, where the loads were distributed, while the Prince and his companions were taking leave of Cagni

Chapter III.

and endeavouring to cheer up his spirits with the hope of overtaking them. By 8.30 the porters had their loads on their heads, and started on their way in a long file, with deafening shouts, on the wide and even road to Kampala. The caravan

THE NATIVE PATH.

numbered about 400 individuals, and the vanguard was nearly out of sight by the time that the Prince and the other members of the expedition started in their turn.

Soon after leaving Entebbe the road enters under the majestic vaults of a tropical forest. The distance from Entebbe to Fort Portal is about 180 miles, with an ascent of some 1,165 feet. This ascent may be regarded as falling into four sections belonging to separate river systems. The first of these collects the waters which flow southward into the River Katongo, a tributary of the Victoria Nyanza. Lake Isolt belongs to this section. The second and third basins contain the affluents of the

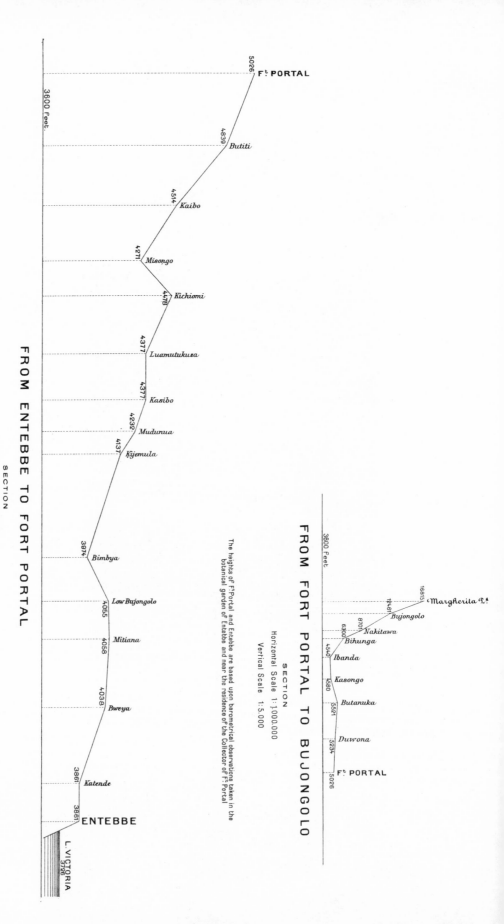

FROM ENTEBBE TO FORT PORTAL

SECTION

Horizontal Scale 1:1.000.000

Vertical Scale 1:5.000

The heights of Fᵗ Portal and Entebbe are based upon barometrical observations taken in the botanical garden of Entebbe and near the residence of the Collector of Fᵗ Portal

3600 Feet.

5026 Fᵗ PORTAL

4839 Butiti

4514 Kaibo

4271 Misongo

4478 Kichiomi

4377 Luamutukuza

4377 Kasibo

4232 Mudurua

4137 Kijemula

3974 Bimbya

4055 Low Bujongolo

4058 Mitiana

4038 Bweya

3861 Katende

3861 ENTEBBE

L. VICTORIA 3726

FROM FORT PORTAL TO BUJONGOLO

SECTION

Horizontal Scale 1:1.000.000

Vertical Scale 1:5.000

The heights of Fᵗ Portal and Entebbe are based upon barometrical observations taken in the botanical garden of Entebbe and near the residence of the Collector of Fᵗ Portal

3600 Feet

16815 Margherita Pᵏ

12461 Bujongolo

8701 Nakitawa

6300 Bihunga

4540 Ibanda

4580 Kasongo

5521 Butanuka

5234 Duwona

5026 Fᵗ PORTAL

From Entebbe to Fort Portal.

Misisi, which takes its course northward and flows into the Albert Nyanza. A last rise brings the traveller to the watershed between the tributaries of Lake Albert Edward and those of Lake Albert. This distribution is brought out in the vertical section annexed.

IN THE TROPICAL FOREST.

This vast region, which forms a sort of tableland between the three lakes, is intersected in every direction by ridges of hills, lower, steeper and more crowded to the east, more distinctly marked into ranges to the west.

71

Chapter III.

The colour of the earth is a brick-red throughout the district. The vegetation is distributed according to accidents of the soil. The high ground, the top of the hills, and their slopes are covered with deep grass and occasional single trees or groups of a few trees set in brushwood. The valley bottoms where water flows are covered with luxuriant forests. Where the waters

BETWEEN SWAMP AND FOREST.

stagnate stretch vast swamps covered with gigantic papyrus, under whose shade flourishes a rich growth of aquatic plants.

From the higher ridges, as far as the eye can reach, stretches an undulating plain, whose rounded hillocks, covered with deep yellow grass, are diversified by low-lying strips of dark green forest.

72

From Entebbe to Fort Portal.

The word *grass*, by the way, is hardly appropriate to a growth which, while reaching a height of from 10 to 20 feet, is at the same time so dense as to make it practically impossible to deviate from the path, and offers far more analogy to a huge bed of reeds than to a meadow. It is called " elephant grass," and is indeed a pasture appropriate to such a herd. From time to time the elephant grass makes way for herbaceous vegetation on a more modest scale, not more than three feet high, and dotted with innumerable flowers.

PLANTAIN GROVES.

The natives are in the habit of setting fire to the grasses during the dry season. Possibly the vast fires thus kindled, and which spread especially on the heights where the earth is dry and where the wind fans the flames, destroy the young trees, and so hinder the formation of forests except in the sheltered valleys beside running water. The fact is that, as a rule, the trees which stand here and there among the tall grasses, and give the country the characteristic look of a park, are all of

73

Chapter III.

very considerable size. The principal species are acacia, mimosa, euphorbia, erithryna, and spathodea, both these last with brilliant scarlet flowers.

At the foot of these trees, among the brushwood and low plants, is nearly always to be found one of those curious cones constructed by termites which characterize all Central African landscapes. It would seem as if some reason must exist for their invariable connection with these isolated groups of trees. Possibly the termites, by collecting earth in one spot, favour the development of bushes and creeping plants, which in their turn serve to shelter the growth of some forest tree until its roots are firmly established in the soil.

The forest zones in the valleys along the brooks are real oases of virgin forest. The luxuriant trees over a hundred feet high, diverse species of acacias, majestic palms (Borassus and Raphia), cassia and dracæna, are overgrown with climbing plants, and entwined with the long ropes of giant lianas. Troops of monkeys are frequently seen leaping from branch to branch with shrill cries. The white-tailed Colobus is the commonest species. The forest soil, even on days of blazing sunshine, remains damp and elastic. Off the path the whole ground is one carpet of deep moss.

The contrast with the open tracts enhances the charm of the forests. After crossing a slope scorched by the sun, the traveller enters into the profound shade heavy with the perfumes of acacia, mimosa, jasmine, and honeysuckle.

The district is fairly populous, but the inhabitants are so hidden away among their banana groves and impenetrable grasses that it is possible to pass quite close to villages without noticing them. They consist of clusters of huts usually situated half-way up a hill, surrounded by tufts of bananas, little cultivated fields and a few forest trees. The huts are of the usual conical type. The circular roof thatched with grass straw is artfully constructed to reach down to the ground on every side except over the entrance, where it is cut short and

BAGANDA WOMEN.

projects into a low narrow porch. The interior is encumbered by the numerous pillars and posts which support this heavy roof. Some of the huts are surrounded by an enclosure, or even by several enclosures, so that three or four courts must be crossed to reach the house.

Chapter III.

The land around the huts is cultivated for a short distance only. As is usual in tropical countries, the indolence of the population limits the production of the soil to the amount which is strictly necessary to sustain life. There is no trace of co-operation. Each family owns its hut and its field, which it

BAGANDA.

cultivates for its own exclusive use. Agricultural labour is performed entirely by women. They cultivate plantain, egg-fruit, potatoes, sweet potatoes, beans, maize, dura, cotton, sesame and sugar-cane. A delicious fruit, always cool and refreshing, is the pawpaw.

From Entebbe to Fort Portal.

The banana, or plantain, is the staple of diet. There are several varieties. Besides the sweet banana, which is eaten ripe and raw, there is a plantain which is gathered unripe and eaten cooked. From the flesh of another variety a sort of bread is made. The juice is pressed out and forms a refreshing, cool

BAGANDA WOMEN.

drink called Mbisi. This becomes alcoholic and intoxicating if allowed to ferment, and is then called Mwenge. Finally the leaves and stalks are used for various purposes. The origin of the cultivated banana is uncertain. Botanically it is quite different from the wild native banana, and it is doubtful whether it could have been derived from it.

Chapter III.

The path, which seems at times like a sort of deep trench cut through walls of high grass, and then again opens out into a tolerably wide road over tracts of plain, proceeds as straight as any ancient Roman highway, crossing hills or following their ridges, descending into valleys and piercing forests, or running over reaches of watery swamps on a low viaduct. This latter is constructed by cutting down papyrus stalks and canes

PAWPAW TREE.

and throwing them across the road from side to side, thus forming a thick stratum upon which the path is built of sand and earth, beaten hard and strengthened on either side by piles driven deep into the mud.

This is the ancient road which existed before the British occupation. It is kept up with great care by gangs of half-naked women, old and young, who weed out the grass and smooth the way with little native spades.

The first part of the road, where the hills are nearer to one another, runs incessantly up and down the steep inclines. After Lake Isolt the slopes become gentler, with intervals of plain, and the marches are consequently less fatiguing. The swamps, too, diminish as the traveller approaches Fort Portal, and the country takes on a healthier look.

PORTERS' HUTS.

The temperature is pleasant in the early morning, but towards midday it becomes very hot, although, fortunately, during the hottest hours the sky is nearly always covered with clouds, which, be they thick or thin, are always sufficient to veil the blazing rays of the sun. Nearly every day or night there is a violent but brief thunderstorm with a gale and torrents of rain. Happily, the Italian caravan had no experience of the terrible storms, accompanied by water-spouts, cyclones and dangerous electrical discharges, which

Chapter III.

inundate, tear up, and destroy everything upon their path, and are said to be not infrequent in Uganda.

The changes of weather are, as a rule, sudden. In a few minutes the sky, up till then clear or scarcely veiled with a light cloud, grows black as midnight and threatening. With equal rapidity, after a brief period of rain the heavy dark clouds are dispelled by the blazing sun.

BUILDING A HUT.

The duration of the marches was from three to six hours, during which period from 10 to 18 miles were covered. The porters, as a rule, walk fast; in some places they nearly run. The caravan usually started with the dawn, about 5.30; and stopped by midday so as to rest during the hot hours. On the march the caravan covered nearly half a kilometre.

WOODED VALLEY BETWEEN ENTEBBE AND FORT PORTAL

From Entebbe to Fort Portal.

The din of the chattering, laughing and shouting was a little diminished in the hard bits of road only where a steep up-hill would set even those who were not loaded panting. From every little village along the way the natives ran out, curious to see the sight and exchange chatter and laughter with the porters. Now and then the travellers met a caravan bringing salt from Toro, or ivory from the Congo, or even a white trader travelling with his own escort.

The native escort exercised a certain discipline over the numerous party, and intervened from time to time to adjust quarrels and disputes started, as a rule, by the porters who, in order to lighten their own labour, would requisition by force any other natives whom they might meet on the road.

The blacks are on the whole childlike, good-natured and peaceable, or ill-tempered and savage, according as they are managed. With a little tact and goodwill, not without necessary firmness, it is easy to direct their impulsive natures.

The great majority of the caravan consisted of Baganda, the real native population of Uganda, whose anthropological characteristics are so diverse as to presuppose the product of mingled elements. Some of their features are distinctly negroid; as, for instance, woolly, jet black hair; the nose sunk at the root, flat and wide; broad, protuberant lips and projecting ears. But the prognathism is not marked, and the brow is wide and not retreating. They are usually lean, not muscular, and do not give the impression of a very strong people.

Their manners and customs seem more advanced than in many other African tribes. They neither dye nor grease their skin; they do not tattoo their persons nor cover themselves with decorative scars, and with the exception of the children

Chapter III.

and a few women they are not loaded with necklaces and bracelets.

Many a traveller has been astonished by their complex social order, a veritable feudal system, while in their legends and traditions, in the designs of their household utensils made of plaited grass, in the form of their musical string instruments, in their astronomical symbols carved upon horns, and in certain burial rites, indications have been suggested of relations and contact with ancient Egypt.

VISIT OF A CHIEF WITH HIS COURT, BRINGING PRESENTS.

The Baganda have preserved the history of their ancient dynasty by pure verbal tradition. It consists of thirty-six names of kings, and must date back as far as the fourteenth or fifteenth century.

The Suahili porters formed a smaller part of the caravan than the Baganda. The Suahili are a cross between Arabs and Bantu negroes. Coming originally from the coast, they are now scattered over the whole of Central Africa.

From Entebbe to Fort Portal.

The encampments were always situated at a certain distance from the villages, in places selected beforehand and prepared for the purpose. There was usually a hut where meals were provided for the Europeans, and one or two sheds to shelter the equipment from the weather. Around the sheds stood the European tents. Mr. Martin formed a second smaller camp. English and Italian flags were flown over each. The tents were surrounded by a zeriba or enclosure of plaited cane

DANCING IN THE PORTERS' CAMP.

which served less as a defence than as a means of dividing the European camp from that of the native porters.

The native porters would arrive at their destination at a run, singing and shouting, throw down their loads hastily on the spot fixed for the purpose, and immediately set to work to build huts for their own shelter. The huts would spring up all around with the most marvellous rapidity. The method of

Chapter III.

construction is most ingenious. A number of slight rods or flexible canes are stuck into the earth in a circle. Their upper ends are bent so as to meet in the middle and interwoven so as to form a dome. Upon this are placed bundles of grass disposed in such a manner as to leave a narrow opening for the entrance. Thus in less than a quarter of an hour a vast grassy plain is transformed into a considerable village. While the work proceeds, there arrive from the neighbouring villages long files of women and old men carrying on their heads parcels of

WRESTLING MATCH AMONG THE PORTERS.

fruit and of sweet potatoes wrapped in plantain leaves. Swarms of naked children accompany them. The tiny ones are carried on their mothers' backs in a fold of their garment. In this way the caravan lives entirely on food supplied by the population of the regions crossed.

In the meantime the Duke would receive a visit of ceremony from some chieftain, whose arrival in camp, attended with the pomp befitting his dignity, had been heralded by

PAPYRUS SWAMP BETWEEN ENTEBBE AND FORT PORTAL

From Entebbe to Fort Portal.

Groups of native converts, strangely contrasting with their heathen surroundings, would pray in a loud voice, or recite the rosary. In addition to the rosary, they were often adorned with

BETWEEN ENTEBBE AND FORT PORTAL.

crosses, medals and reliquaries hanging around their necks. Here a Mohammedan would be worshipping on his bit of carpet at sunset, while yonder the native escort would be going through their daily drill.

As evening closes in, the camp is lit up by hundreds of fires, around which the porters sit until far on in the night, roasting the sweet potatoes, or boiling the plantains which, with the addition occasionally of a little dried fish, form their sole diet.

After a march of five or six hours over heavy ground, carrying fairly substantial loads on their heads, this frugal single meal was amply sufficient to their simple needs. Banana wine is a rare luxury, while water is scarce and filthy, with a disgusting smell and taste even when boiled.

Chapter III.

The various members of the expedition were by no means idle during camp. The Duke was in the habit of attending in person to the sorting and verifying of the equipment; to the meteorological observations taken with instruments arranged in the little camp observatory which was set up at each camp in the most suitable position; and to the observations of longitude and latitude.

CAMP AT BUJONGO.

At each halt Dr. Cavalli was immediately besieged by crowds of native patients from every district, while one or another of the porters was sure to make daily demands upon his treatment.

Sella, besides his photographic work, would spend part of the afternoon in roaming in the neighbourhood of the camp with Roccati and Cavalli in quest of botanical and zoological specimens. Contrary to expectation, the lower forms of animal life proved to be rare. Possibly they have been annihilated by the termites which invade and destroy everything. Native men and boys from the neighbouring villages would join eagerly in the quest, and show visible amusement at the sight of Roccati treasuring up diminutive insects, spiders and scorpions, and putting by lizards and chamæleons.

CAMP AT KATENDE.

Now and again a shooting party would set forth. Guinea-fowls and doves abound in the plantations around the camps. The region is rich in elephants, zebras, antelopes, lions and leopards. This sort of game, however, requires special beating, and is not compatible with the rapid marches of a caravan bent upon a totally different aim. It was only very seldom and

Chapter III.

at a great distance that an occasional antelope was seen fleeing from the approach of the party.

Towards evening the air would grow cool, and after dinner the Europeans would gather round the now far from unpleasant warmth of a great blazing fire which served also as protection against mosquitoes. At night the latter became a real plague; through long and wakeful hours you would hear their drone diversified by the cry of the sentinels and the strange trill of the *buga-buga*, a tiny bird which builds its nest around camping places.

Between three and four in the morning the réveillé was sounded by trumpets and the camp at once filled with din. It took the caravan little more than an hour to get ready to start. The porters fell upon their loads and set forth with their usual shrill cries.

The journey from Entebbe to Fort Portal took fifteen days.

On the 18th of May they encamped at Mitiana, near a branch of the French Roman Catholic Mission at the foot of a hill, on the top of which stood a small shrine. They exchanged visits with the Missionary Fathers, who sent presents of excellent European fruits and vegetables. On the following night they reached Bujongo in sight of Lake Isolt, a lake rich in fish and dotted with wooded islands. This and the following camps were fortified with strong hedges and palisades, because the country was infested with lions to such an extent that the natives would not risk travelling by night.

On the 25th of May they crossed the border between the Province of Uganda and the Western Province, which includes the districts of Toro, Unyoro and Ankole. In addition to the native chieftains of the new district, followed by their respective courts, the Prince was here met by the Sub-Commissioner of the province, Mr. A. F. Knowles, who was to accompany the expedition throughout his own jurisdiction, while Mr. Martin returned with his escort to Entebbe.

RUWENZORI SEEN FROM BUTITI.

Henceforward the réveillé was no longer sounded by trumpets, but by the rolling of the Unyoro drums. The game now seemed to become more abundant. Numerous deep elephant tracks crossed the path. Herds of antelopes became more frequently visible in the far distance. Vultures, hawks, and other birds of prey wheeled in the sky.

A new feature of the landscape was the granite formation, which here and there pushes its way through the soil in rounded hummocks somewhat similar to the rocks known as *moutonnées*, in regions which have passed through a glacial period. The grass became less deep, the trees and flowering shrubs increased

Chapter III.

in numbers, while between the hillocks were open spaces of ground nearly barren save for a growth of reddish-yellow grass mingled with low ferns. The plantain groves diminished in extent and were in part replaced by sweet potatoes and beans. The district was less thickly populated than that which preceded.

The march was often heavy and fatiguing. The weather had changed for the worse, and frequent rainfalls made the track muddy and slippery.

As the expedition drew nearer to the Lakes Albert and Albert Edward, their impatience to see the chain of Ruwenzori

NEAR BUTITI, WITH RUWENZORI IN THE BACKGROUND.

grew acute, and from the moment when they entered the Toro district their attention concentrated itself upon the western horizon, especially when the path led them over the top of some

Baker · Stanley · Speke

VIEW OF RUWENZORI FROM A HILL NEAR KAIBO

From Entebbe to Fort Portal.

hill. Twice they fancied that they had sight of snowy peaks, but it was an illusion created by white clouds upon the horizon.

Finally, on the morning of the 28th of May, from the top of certain hills to the north of Kaibo, which form part of the watershed between Lake Albert Edward and Lake Albert, on a day when the sky was clouded but the air clear, they suddenly saw against the sky to the westward the snowy peaks of the great chain, which were about 45 miles off, and looked as if they were suspended in the air, for their feet were enveloped in mists while a heavy rack of clouds hung so low over the summits as nearly to rest upon them. Seen from this point, the mountains appear divided into three main groups. Of these, the central one is dominated by a characteristically cloven peak, covered with snow, which seems to be the highest of all, and which is separated from the group to its south by a deep notch; the third group is to the north or north-east of the central mass. The foot of the glaciers, which come down from the high ridges, is hidden by the projecting spurs of the range.

They encamped that day at Butiti, where the Missions (Protestant and Roman Catholic) were abundantly hospitable. The camp was protected by a strong palisade guarded by sentinels, and great fires were kindled all around. Occasional roars heard distinctly through the silence of the night showed that these precautions were far from superfluous. Two weeks later, at Misonga, not far from Butiti, a lion made its way into Cagni's camp and succeeded in escaping unharmed, thanks to the darkness of the night.

On the following morning, May 29th, about an hour and a-half from Butiti, the Prince and his companions climbed a hill close to the path in order to get another look at Ruwenzori,

Chapter III.

which was here visible in all its splendour. They had now come further north-west, and hence the northernmost group of the chain appeared nearer to the central group, which from this point also appeared to be the highest of all, and to show the greatest extent of glacier.

The sky was clear over the mountains to westward, but dark and stormy in the east. Right and left stretched an undulating plain with low rounded hills, reddish or earthy yellow, dotted with dark green patches of euphorbia, or of the light and finely cut foliage of the acacia. Farther off, the landscape melted into the misty distance, and finally vanished from sight near the foot of the mighty spurs of the range.

Moore was reminded in these regions of the Alps as seen from the Piedmontese or Lombard Plains, but the comparison does not hold good. The difference is profound, although so subtle as to baffle analysis. It is true that the far-off slopes clad with elephant grass, and the swamps hidden under tufted papyrus resemble our hills and our cultivated valleys. There is no definite sign to indicate that those far-off plains, which to all appearance might consist of meadows and cornfields, maize plantations and orchards, are in reality the lair of elephants, buffaloes, antelopes and lions. Yet still the picture is in a different key, with a grim solemnity of its own. The likeness is the fruit rather of a mental comparison than of a real, direct impression from Nature. Signs of the handiwork of man are nearly totally absent. The huts of the natives, their banana groves and their simple crops are only just visible on closer inspection of the landscape, of which they form an insignificant detail, hardly touching its virgin and primitive aspect.

A little further on the party crossed their last forest, the finest of all that had lain across their path, and swarming with

94

FOREST BETWEEN BUTITI AND FORT PORTAL.

TROPICAL FOREST NEAR FORT PORTAL

monkeys. After a long march of seven hours they reached Fort Portal on the same day.

On approaching Fort Portal, H.R.H. was received by the King of Toro, Kasagama, a handsome man, above the average stature, with an open and intelligent countenance. He was accompanied by a large escort carrying numerous gifts. After crossing the belt of cultivated fields around the city, they entered the wide, clean street of Toro blazing with sun, and flanked on either side by the chiefs of the region who had come with escorts to receive the expedition. The street was crammed with people, and especially with noisy children.

Fort Portal was founded in 1891 by Capt. Lugard after he had deposed Kabarega, who was an ally of Mwanga in his revolt, and had set Kasagama upon the throne in his place, thus ending a period of frightful persecutions and raids which had nearly depopulated the country. The site of Fort Portal is very healthy. It stands at a height of 5,000 feet above the sea, in a basin bounded to the west by the range of Ruwenzori, which slopes down towards Lake Albert in a long chain of gradually lessening spurs, and to the east by the hills which divide the basin of Lake Albert Edward from that of Lake Albert. Of the great range only the highest points are visible, when by exception they are not covered with clouds, above an advanced buttress known as the Portal Peaks. To the north-west, at the foot of the mountains, are scattered volcanic cones among which lie numerous small crater lakes.

The European residents of Fort Portal, including ladies, are scarcely fifteen in number. They consist of the Sub-Commissioner, the Collector, the Commander of the troops, and the Catholic and Protestant Missionaries. The dwellings

Chapter III.

of the English officials, including the residence of the Sub-Commissioner, surrounded by a hedge and a palisade, stand upon a hill. Upon the neighbouring hills are the Missions and the Hospital. Upon another hill to the south-east, covered with extensive banana plantations, are situated the dwellings of the King of Toro. On the low ground

THE CARAVAN ON THE MARCH.

between the hills stand the shops, like those of Kampala, in long lines on either side of a wide avenue planted with trees. There are also barracks for native troops, and the usual market. There are many natives here of the Bahima tribe. These are handsome people, alleged to be of Ethiopian origin, tall of stature, slender of figure, with finely proportioned limbs, a somewhat lighter colour than the Baganda, and regular features similar to those of the white races. They are all shepherds; they wear a cloak of skins, and speak a language of their own. The pure type is growing rare on account of their mingling with the Baganda tribes.

The expedition was hospitably entertained in Fort Portal at the residence of Mr. J. O. Haldane, the Collector. The porters encamped on the low ground at the foot of the hill.

CHAPTER IV.

From Fort Portal to Bujongolo—Mobuku Valley.

Two days at Fort Portal—Meeting with Dr. Wollaston—Hesitation about the Route—Departure from Fort Portal—Duwona—Ford of the Wimi River—Kasongo—The Peaks of Ruwenzori once more in sight—Entrance into the Mobuku Valley — Ibanda — The Duwoni of Johnston — Bihunga — Mahoma Valley — The Moraine of Nakitawa — Discovery of the Bujuku Valley—Bakonjo Porters—Crossing the Swamp—Kichuchu—The Heath Forest—The Flowery Plain of Buamba—Bujongolo—An Icy Night on the Equator.

THE expedition spent two whole days at Fort Portal with bad weather and clouded sky. In spite of these unfavourable conditions, the Duke was able to complete some astronomical observations. An intermediate meteorological station was established at Fort Portal. The observations taken here were to be compared later on both with those subsequently to be taken at Entebbe, on Lake Victoria, and among the mountains, in the valleys and on the summits, in order to supply full data for an exact calculation of altitudes.

At Fort Portal, the party had the pleasure of making the acquaintance of the Rev. A. B. Fisher and of Mrs. Fisher, who had twice ascended the Mobuku Valley as far as the glacier. Another interesting and pleasant acquaintance was that of the Alpine climber, Mr. A. F. Wollaston, who had left the British Museum Expedition for a few days upon the invitation of the Sub-Commissioner, Mr. Knowles, and had come down to Fort Portal on purpose to meet the Prince.

ON THE PUBLIC SQUARE, FORT PORTAL.

As was mentioned in Chapter I, in the months immediately preceding the arrival of the Italian Expedition, Wollaston had made the ascent of some of the peaks at the head of the Mobuku Valley, from the top of which he had made out through the mist two other snowy summits to the north-east, higher than those which he had ascended and seeming to rise above the western slopes of the chain towards the Congo. He had not been able to make out whether these higher mountains were connected with the peaks of the Mobuku Valley.

101

Chapter IV.

The mountains seen and drawn by Stuhlmann at the head of the Butagu Valley to the west of the chain would not, in this case, have been the same as those seen from the east, which the Italian expedition had observed from Kaibo and Butiti. It seemed, therefore, a better plan to attempt the ascent from the western slope.

These accounts perplexed the Duke greatly. If, on the one hand, he followed the route of his predecessors up the Mobuku Valley there was the risk, on reaching the peaks at the head of the valley, of seeing his way to the higher summits cut off by some deep valley or insuperable ridge. If, on the other hand, he should resolve to try the western slopes, it would become necessary to make a long détour across the low regions, through malarial districts, in order to turn the southern end of the

NATIVE HUT.

chain and reach the Semliki Valley. Here, there would be uncertainty as to the sufficiency of local resources to feed so numerous an expedition, and still greater uncertainty as to the disposition of the natives, who were known to be frequently hostile and turbulent in the Congo District.

Of the two alternatives the latter seemed certainly to offer the more serious risks. The Duke of the Abruzzi decided

MARKET, FORT PORTAL.

therefore to follow the more direct and shorter route, ascending the Mobuku Valley and arriving comparatively quickly among the high mountains, where it would be possible to obtain data for forming a decision as to the future route.

The two days at Fort Portal had been days of complete idleness for the native porters, and had been sufficient to undermine and break up the discipline to which they had become accustomed during their two weeks of steady work. When the drum and the trumpets sounded the réveillé at 4.30 a.m., June 1st, not one of the whole troop was ready. Boys and porters dropped in late, one by one, and it took over two hours to get the caravan into marching order. At

103

Chapter IV.

last it started, with the usual shouts, preceded by the English and Italian flags.

The baggage was already diminished by the rations consumed during the preceding fortnight. It was now further reduced by a number of personal effects which were left behind at Fort Portal. Consequently, a portion of the porters had been dismissed, and those retained were selected among the strongest and healthiest.

The Prince was accompanied on his start from Fort Portal by Mr. Knowles, the Collector, Mr. Haldane and Mr. Wollaston,

HILLS NEAR FORT PORTAL.

who was on his way to rejoin the British Museum Expedition in the Nyamwamba Valley.

An escort of twenty native soldiers accompanied the caravan. Their wives had come to bid them farewell. The form of their

From Fort Portal to Bujongolo—Mobuku Valley.

leave-taking was as sober and dignified as possible : each woman knelt before her husband, who placed one hand upon her head.

As has been said, Fort Portal is situated upon the heights which divide the basin of Lake Albert from that of Lake

FORT PORTAL.

Albert Edward. The latter is connected by a short, narrow watercourse with Lake Dueru or Ruisamba, which lies in the hollow called " Albertine Valley," at the foot of the eastern slope of Ruwenzori, and receives all the waters which flow down from the chain on that side. To reach the Mobuku Valley the path skirts the upper basin of Lake Ruisamba without descending to the lake, running first along the hills which form the watershed, and then following the eastern foot of the

105

Chapter IV.

chain from north to south, and crossing the lower course of the valleys and torrents which run down from the ridges.

The country is fertile and well-watered, but very sparsely cultivated excepting in the neighbourhood of Fort Portal. The population is wretched and unhealthy looking. The path, now a mere track, now widening out into a road, is in many places a true mountain trail, which would be extremely fatiguing and even difficult were it not kept up with great care. Natives are to be met at every step, especially women and old men, employed in mending and weeding it. The women, as usual, carry their babies on their back or at their breast and keep their larger children by them. Children and adults are absolutely naked, or else wear rags or skins around the loins. The women adorn themselves with bracelets or, lacking these, tie rings of twisted banana leaves round their arms and ankles.

The way between Fort Portal and the Mobuku Valley was traversed in three stages. After leaving the European station the path first descends over the wide road of the Mpango Valley and crosses the river on a wooden bridge. Next, it ascends to the Royal Hill, where King Kasagama, surrounded by his whole court, waited for H.R.H. Another brief halt was made at Notre-Dame de la Neige to take leave of the courteous Fathers of the French Mission. Tall hedges run on either side of the path, which winds between numerous huts scattered in fields of pease, millet, sweet potatoes and tobacco, and extensive plantain groves.

The way now led south-west, making straight for the mountains. Low hills were crossed by easy slopes, and four hours brought the expedition to Duwona camp, which stands against the foot of the mountain among blossoming euphorbia trees.   Below lies the Albertine Valley, dotted with small volcanic cones.   The peaks above were shrouded in dark mist. The rest of the sky was clear, and the day ended in a limpid sunset.

On the following morning, the way struck southward, first skirting wide low ridges covered with elephant grass ; and then

KING KASAGAMA AND HIS COURT.

crossing by steep ascents and descents the foot of divers spurs of the chain.   The way skirts the mountain so closely that the snowy peaks are hidden from sight.   Numerous torrents had to be forded.   Only one of these was of a certain size, namely, the Wimi, which, when swollen, may become a serious obstacle.   The expedition found it about 30 feet wide, the

107

water very cold, from two to three feet deep, and the current fairly swift. A line of men was formed in the water, stretching from one bank to the other, and the porters with their loads crossed up-stream of them. In this way any man who slipped

NEAR FORT PORTAL.

or staggered was immediately caught and held; and in the space of about one hour the whole caravan was gathered on the opposite bank, which was very steep and covered with thick grass. Not a single parcel had been lost.

From Fort Portal to Bujongolo—Mobuku Valley.

The camp of Kasongo was reached before noon. This camp stands high upon one of the spurs of the range. Lake Ruisamba was just visible through the mists which hid the plain.

Between Kasongo and the Mobuku Valley there was still one last valley to be crossed, known as the Hima.

Soon after leaving the camp, on the morning of the 3rd of June, a portion of the high chain appeared in sight to the westward, framed between the sides of the valleys. First appeared two rocky peaks* with a great glacier at their feet. As the expedition proceeded southward, and went down into the Valley of Hima, these peaks were gradually hidden ; while to their right, that is northward of them, came into sight, bit by bit, the double peak† covered with snow, which, as seen from Kaibo and Butiti, appeared to form part of the central group, and to be the highest of all.

The Hima River was crossed upon a light bridge, and after this the path ran up the valley for a short distance westward, then turned southward again and ascended the slope of the buttress, behind which lay the Mobuku Valley. It was still early in the morning when the expedition reached the top of the ridge, and commenced to descend the other slope into the Mobuku Valley.

Meantime the peaks of Ruwenzori continued to come out one after another to the westward. To the right of the double snow peak, and separated from it by a low, wide col, appeared another group of peaks,‡ which extended northward in the shape of an ice ridge edged by a big cornice, under which stretched a glacier.

---

* Elena and Savoia Peaks of the map.
† Alexandra and Margherita Peaks.
‡ Mt. Speke (*see* illustrations, pp. 115–116).

Chapter IV.

Thus, as the caravan had proceeded from north to south, the peaks of the chain had become visible in inverse order from south to north. In this way two rocky peaks had come into view, connected by a wide glacier with the twin peaks covered with snow. These four together formed what from Kaibo and Butiti appeared as the central group of the chain.* Next had followed a wide depression, after which the ridge had risen again and formed two great peaks of rock and ice which

WOMEN AT CAMP DUWONA.

stretched northward so as to form a long snowy crest. Only this last group, which was, without doubt, the Duwoni of Sir Harry Johnston, was visible from the Mobuku Valley.

Mt. Stanley.

110

From Fort Portal to Bujongolo—Mobuku Valley.

The path now went down to the Mobuku River, which flows in a bed about 25 feet deep hollowed out through ancient alluvial deposits. This torrent is some 60 feet wide, the

ACACIAS ON THE ROAD BETWEEN DUWONA AND KASONGO.

water nearly three feet deep, and the current violent. The water is cool, but of a yellowish hue, which does not make it attractive to drink.

While the caravan was collecting on the bank, the chiefs of the neighbouring villages were arriving from either side of the valley, with their attendants bearing stools and umbrellas and followed by troops of natives. They all took part in helping the caravan to cross the ford. A rope was stretched across the current, and numbers of natives took up their

111

Chapter IV.

positions below the rope to give greater security. The porters, with their loads, straggled across up-stream of the cord and holding by it. In this way the whole party was soon reassembled on the other bank of the Mobuku without accident and resumed their way, now ascending the wide level valley bottom as far as the camp of Ibanda.

Ibanda (4,540 feet) stands upon the right bank of the Mobuku River at a point where the valley widens into a plain more than one mile wide, shut in by rounded hills and covered

BETWEEN DUWONA AND KASONGO.

with deep grasses and a few scattered trees. A small tributary valley opens near the camp. Further up, the valley appears to be completely shut in by a high and steep peak which

From Fort Portal to Bujongolo—Mobuku Valley.

forms one of the Portals. Beyond this, again, rises the snowy mountain which has been already described, and which Sir Harry Johnston had named Duwoni.

The general trend of the valley is from east to west. Marks of glacial action are evident. A little above Ibanda, on the opposite side of the valley, lies a stretch of marginal moraine about thirty yards deep. A number of spurs seem to be the remains of frontal moraines cut off by the torrent. There are numerous boulders and round smooth rocks of the type known as *moutonnées*. Finally, looking down the valley, a transversal ridge has every appearance of a terminal moraine.

Round about the camp are numerous villages and plantain groves. The natives are naked, with strings of shells round their loins from which bits of cloth are suspended.

At Ibanda there is not the usual shed for eating under cover. Fortunately the weather was fine, and a few trees near to the torrent offered shade for the midday meal. Not a single fish was to be found, in spite of long and attentive inspection of the water.

The evening was perfectly clear and the light died away slowly. The familiar sound of the torrent called to memory quiet evenings passed in some remote valley of our own Alps. Below the camp blazed numerous fires which now and again seemed extinguished and rekindled as the dark shapes of the natives flitted busily to and fro in front of them. The mountain walls of the valley stood out clear on the starry sky. The snows of Duwoni glittered softly in the bright starlight.

The prospect seemed very hopeful. The Italian expedition were more fortunate than their predecessors in the circumstance that, before even reaching the feet of the mountains, they had sight of many peaks, and were able to ascertain the important

Chapter IV.

fact that the double peak seen from Kaibo and Butiti as in the centre of the chain and appearing to be the highest of all, is not the Duwoni of Johnston.

FORD OF WIMI RIVER.

Leaving the camp of Ibanda the march followed rapidly across the high plain, which was swampy here and there, with groves of tall acacia and dracæna and dotted with round smooth boulders. Soon they reached the foot of another buttress, a spur of the right-hand slope of the valley. Here the path became so steep at some points that even those who had no load to carry got out of breath. The natives, who during the first part of the stage kept up their usual cheerful hubbub, now became silent as they panted up the wearisome ascent, and scattered far and wide, covering a long reach of the way.

114

From Fort Portal to Bujongolo—Mobuku Valley.

As the valley rises it puts on little by little a grim and mysterious aspect. The forbidding precipices of the Portal peaks seem absolutely to close its deep western recesses.

About half-way up the spur is a narrow grassy ledge, where are perched a few native huts. These are the last human habitations of this valley. Beyond everything is desert. The place is called Bihunga, 1,760 feet above Ibanda, and 6,300 feet above the sea. Here the British Museum Expedition had spent

Alexandra and Margherita Peaks.   Mt. Speke.

THE SNOW PEAKS OF RUWENZORI, SEEN FROM THE HIMA VALLEY.

several months in collecting material for research. A spacious hut still stood as a record of its sojourn.

The tents were pitched around this hut with difficulty,

owing to the small space of level ground available. The porters encamped as best they could on the steep slope.

The view of the mountains was entirely cut off by the spur upon which the camp stood. On the other side they overlooked

MT. SPEKE (THE DUWONI OF JOHNSTON) SEEN FROM THE LOWER MOBUKU VALLEY.

the plain of Ibanda, and down the wide valley till the point where everything disappeared in the misty atmosphere. The near hill sides were clad with dense forest diversified by small clearings covered with deep grass. There was scarcely any sign of animal life. Near the camp they saw lobelias for the first time. There were many dracænas, and a most beautiful erithryna covered with flame-coloured blossoms. A narrow strip of small cultivated fields surrounds the tiny village, which is inhabited by a few Bakonjos, naked in spite of the cold of this high region.

116

From Fort Portal to Bujongolo—Mobuku Valley.

At Bihunga the Duke began to reduce the number of his caravan. Henceforward the way was to lead through uninhabited regions where the commissariat would offer increasing difficulties. An agreement was entered into with the chiefs of the villages around Ibanda by which they were to send parties of porters regularly up the valley with provisions. In addition to the limited nature of the resources of so small a district, the actual distance to cross and the difficulty of the marches would increase as the expedition proceeded upwards.

FORD OF MOBUKU RIVER.

They left behind at Bihunga a portion of the baggage, consisting of some forty loads, including all those personal

Chapter IV.

effects which became unnecessary in the cold climate of the high mountains, and a number of the boys were also dispensed with. These, as well as the superfluous porters, went back to Butanuka, a village half-way between the Mobuku Valley and

FORD OF MOBUKU RIVER.

Fort Portal, which now became the halting-place for all the Baganda who were sent back from the mountains. Last, but not least, the twenty native soldiers of the escort, commanded by Sergeant Green, remained at Bihunga, where they formed a link between the expedition and the lower valley. The portion of the supplies and equipment which was left behind was sheltered in the hut of the British Museum Expedition.

On the morning of the 5th of June, the caravan again set out upon its way. An extremely narrow and very steep path through thick brushwood and thorny branches, which scratched the face and hands of the travellers, led up the spur of Bihunga, and then crossed the tiny Chawa Valley and redescended into the Valley of Mahoma, an important tributary on the right hand of the Mobuku.

The descent was steep, through a dense forest of tall trees which climbed high up on the precipitous sides of the valley. Numerous specimens of a fine conifer, the podocarpus, were overgrown with a tangle of creeping plants diversified with brilliant orchids. Under the trees was a dense leafy under-growth mingled with ferns of numerous species, forming so impenetrable a brushwood that the path became a veritable tunnel, where one had to walk bent double for long tracts. The bushes and creeping plants covered many fallen tree-trunks, from the rich soil under which numerous specimens were added to the zoological collections. The ground was very damp, in many places soaking, and extremely slippery, and the porters had difficulty in keeping their feet. The way ran through the forest as far as the banks of the Mahoma.

Once the torrent crossed, the path wound among ferns and tree-ferns of several varieties up a slope so steep as to be extremely laborious for the porters, who marched disbanded and very slowly. At a certain point of altitude the first bamboos and heaths appeared among the ferns. The ground was slippery and muddy, and scattered with rocks of every dimension.

This slope is merely a great lateral moraine of the glacier which once flowed down the valley and probably covered the whole plain of Ibanda. It is unaccountable that the real nature of this ridge should have escaped the notice of so many previous explorers of the Mobuku Valley. A corre-sponding and parallel moraine runs along the opposite or left-hand side of the valley. The Mobuku torrent roars more than 600 feet below in the deep and precipitous gorge where it has cut a channel through the detritus, while the blocks and pebbles of the moraine formation are quite plainly visible in section on the deeply cut sides.

The trees now grew denser and denser until, on the top of the moraine, the path once more entered the forest. For some distance the way followed the crest of the moraine, in many places less than a yard wide, until it reached a gigantic boulder of gneiss about 30 feet long, and from 18 to 20 feet high, near which stood a little straw-roofed shed quite crooked and propped up by a few piles driven into the earth. This

IBANDA.

is the camp of Nakitawa. On every side stretched the forest of tall trees with the dense brushwood beneath. Hours of hard work were required to cut down enough

From Fort Portal to Bujongolo—Mobuku Valley.

brushwood and trees to make room for the seven tents. At the foot of the boulder the natives crowded around the kitchen place. This camp is 8,700 feet above the sea-level. In

THE PORTAL PEAKS ON THE WAY UP TO BIHUNGA.

spite of occasional descents a rise of 2,400 feet had been accomplished in one march. During the whole afternoon the porters kept dropping in, one by one, tired out with the hard day's journey. The Baganda are a people of the plains, and evidently incapable of enduring the fatigue of mountain marches. It had now become obviously necessary to replace them by Bakonjo, who are acclimatized to this valley and accustomed to climb its slopes in the chase after marmots and hyrax.

Every slope in sight was covered by the forest. It was a scene of virgin and untouched Nature. The regions inhabited by man had been indeed left behind.

Chapter IV.

Near Nakitawa, at the entrance of the Mahoma Valley, the ancient moraines of the two valleys meet and unite together. In the corner formed by the meeting of the left moraine of Mahoma with the right moraine of Mobuku lies a little lake, which was visited subsequently by the expedition on their return journey.

The peaks of the Portal group soar up over the left side of the valley exactly opposite this camp. The two southernmost of these peaks stand like giant sentries on either side of the entrance of another great valley which here opens into the Mobuku.

BUILDING SHEDS, IBANDA.

The discovery of this important tributary valley, over-looked by all the previous explorers, permitted the Duke to arrive from the first at certain vital conclusions regarding the position of the peaks.

From Fort Portal to Bujongolo—Mobuku Valley.

It was, as a matter of fact, through the opening of this new valley and above its head that the expedition had seen the Duwoni of Johnston from Ibanda. In consequence it became evident that this mountain does not stand at the

HILLSIDE, BELOW BIHUNGA.

head of the Mobuku Valley. Furthermore, comparing the aspect of the chain as seen from Kaibo and Butiti with the successive sight of the single peaks, as descried in crossing the Hima Valley between Kasongo and Ibanda, and subsequently in descending into the Mobuku Valley, it had become quite plain that the peaks and glaciers of the highest central group

123

Chapter IV.

were to the south of Duwoni. Consequently the whole of this group must stand between Duwoni and the Mobuku Valley.

Hence it seemed obvious that the newly discovered valley must lead into the very heart of the chain and penetrate amongst its highest peaks far more directly than the Mobuku Valley.

Owing, however, to the absence of all accounts of this valley and the uncertainty as to whether it was accessible up to the foot of the mountains, H.R.H. decided to continue on the road followed by preceding explorers, so as to lose no time in reaching some high point whence he might be able to judge of the relative positions of the peaks and valleys.

The Duke of the Abruzzi preserved the name of Bujuku for the newly discovered valley, this being the name by which it was known to the inhabitants of Ibanda.

For five successive days the weather had been unusually fine for these regions, nor were they again to enjoy so long a period of uninterrupted clear sky during the whole campaign, except quite at the last when they were on the point of leaving the mountains. On the morning of the 6th of June, with the dawn, a fine rain was falling from the grey cloudy sky.

Provisions were expected by porters, who only arrived at about seven o'clock and consisted of eighty Bakonjos. These are tall men of robust habit, with somewhat prominent jaw, their hair is either shaven or disposed in strange fashion, and they frequently wear a small beard. Their skin is tanned by the sun, the rain, and the cold, and is hard and rough as leather. They wear a piece of cloth hanging from the loins, bracelets of metal or cord round their arms and legs, and a fur pouch suspended from the neck for pipe and tobacco. Some wear a leopard skin over their shoulders, or a cloak made of rabbit pelts (hyrax) stitched

124

THE HEATH FOREST

together. There are no converts among them. They carry long staves on their march and use them with great skill in the difficult places. These eighty men were now kept to replace half of the Baganda porters who were at once sent down. Everything was set in order. The men were refreshed with food, and at last the expedition started at about eight o'clock.

After Nakitawa the path, now reduced to a mere trail, descends from the brow of the moraine, skirting its slope

BIHUNGA.

through bamboos and creeping plants, to the bottom of the valley which here opens out into a plain. The way now leads across this terrace to the Mobuku torrent, here so small that it can be crossed dry-shod, leaping from stone to stone. A tree trunk thrown across the stream made the passage easier for the porters.

The enormous difference in the volume of the Mobuku River at Ibanda and above Nakitawa must be specially due to the

FOREST ABOVE BIHUNGA.

influx of the Bujuku River below the latter point, and points to the conclusion that the supposed tributary is really the more important of the two rivers.

The flat valley bottom is a lake of mud upon which grows a forest, nearly entirely composed of bamboos. The path is all water and mud. You sink in to the knee. Under the mud the foot meets with stones or pieces of wood, or is caught in a creeper or a fallen trunk, making it necessary to grasp the surrounding bushes, frequently thorny, so as not to lose balance. Little by little you learn to take precautions in walking, to recognize the points likely to afford solid foothold ; to proceed now by jumps and again by placing one foot to the right and the other to the left of the path, perching upon stones or upon roots which rise above the mud or upon fallen branches of trees, or again by preserving

FOREST AT THE MOUTH OF THE MAHOMA.

your equilibrium along a fallen tree-trunk. But, even so, you frequently become entangled or get stuck, and seek solace in expletives which are more energetic and expressive than elegant. Meantime, rain began to fall heavily, and

127

from the bamboos, from the heaths, from the tall ferns, and from all the leafage of the forest, a chilly drip fell ceaselessly upon the travellers.

Bedaubed with mud from head to foot, their clothes

TREE - FERNS.

soaked in water, after crossing the valley as far as its left slope, the expedition reached the foot of a high overhanging cliff at the bottom of a short valley shut in by a moraine. This was the so-called Kichuchu Camp, at a height of 9,833 feet above the sea-level, and 1,133 feet above Nakitawa. The rocky wall forms a shelter over a narrow strip, where you are indeed protected from the pouring rain, but where the soil is soaked with the water which drips off the rock upon it. Here there was room for a single tent only. All

THE MOBUKU RIVER IN THE HEATH FOREST

From Fort Portal to Bujongolo—Mobuku Valley.

around was deep mud. Branches and tree trunks spread upon the ground formed a platform large enough to admit of two more tents. It took many hours' hard labour in the mud and under the rain before the camp could be got ready.

Such firewood as could be collected in the immediate neighbourhood was scarce, and the fires insufficient. The remaining Baganda porters, tired, discouraged, and shivering with cold were evidently incapable of proceeding any further. They were therefore all sent back with the boys to Butanuka, thence to join their comrades who had been dismissed from Bihunga and Nakitawa. Henceforward the expedition proceeded with the Bakonjo only, leaving a number of loads behind to be sent for later as required.

The Kichuchu shelter stands upon a plateau which forms the first of a series of three terraces, all soaked with stagnant water and divided one from another by cliffs some 600 feet to 1,000 feet high. These three terraces form the upper Mobuku Valley. Above Kichuchu the way suddenly grows steep, and mounts by a narrow natural ledge in the rock of a spur about 900 feet high, belonging to the southernmost peak of the Portal group.

At the narrowest and most difficult points of this rocky ledge the climb is facilitated by wooden steps. The path is so steep that you have to climb with hands and feet, clutching the few creeping plants and shrubs which grow within reach. The last bit is less steep, but is again a mass of mud, stones and roots.

The summit is at last reached. This is the brow of the second plateau of the valley, and here one of the most singular sights seen in all the journey awaited the expedition.

129

Chapter IV.

The plateau is completely covered by a great forest of tree heaths. In this forest trunks and boughs are entirely smothered in a thick layer of mosses which hang like waving beards from every spray, cushion and englobe every knot, curl and swell

KICHUCHU.

around each twig, deform every outline and obliterate every feature, till the trees are a mere mass of grotesque contortions, monstrous tumefactions of the discoloured, leprous growth. No leaf is to be seen save on the very top-most twigs, yet the forest is dark owing to the dense network of trunks and branches. The soil disappears altogether under innumerable dead trunks, heaped one upon another in intricate piles, covered with mosses, viscous and slippery where exposed to the air ; black, naked, and yet neither mildewed nor rotten where they have lain for years and years in deep holes. No forest can be grimmer and stranger than this. The vegetation

seems primeval, of some period when forms were uncertain and
provisory. The silence is profound, and the absence of any
sign of life completes the image of a remote age before the
beginning of animal existence, such as might have been those
forests which have given us the strata of coal fossils.

Faint and indistinct tracks on the moss and the fallen
trunks indicate the way. The travellers proceed, leaping and

LOBELIAS IN THE HEATH FOREST.

balancing themselves upon the slippery trunks, in continual
danger of putting their foot in a deep hole and falling in the
openings between the trunks, whence they would be likely to
emerge with broken bones or other injuries. The Bakonjos

Chapter IV.

give proof of marvellous agility. They jump from trunk to trunk; they crouch or crawl to slip their loads under the lower branches; they perform miracles of equilibrium upon sloping trunks, walking all the time so fast that it is difficult to keep up with them.

The path now returns to the Mobuku, which here is a mere Alpine stream buried in the fantastic vegetation on its banks, and roofed over with the strange branches mingling and crossing above it. The yellow-brown waters are without fish or any other form of animal life. The expedition crossed this stream to its right bank, and reached the foot of another ledge, about 600 feet high, formed by an ancient moraine, and likewise covered with heath forest and underwood of tall ferns, creeping plants, orchids and thorny brambles laden with blossoms and with unripe blackberries. In their shade grow violets, ranunculus, geraniums, epilobium, umbelliferous species and thistles. The ledge leads to the third terrace, where there is another rock shelter called Buamba, 11,542 feet above the sea.

Once upon the brow of this ledge and out of the oppressive lifelessness of the heath forest, the expedition found itself suddenly and without transition in the presence of a picture totally different, though no less strange. The long level valley bottom, walled in by towering cliffs on either side, stretched up to the foot of another step, beyond which the valley narrowed into a gorge where stands the shelter of Bujongolo. The peak of Kiyanja* with its glaciers rose far off and high above the head of the valley.

The whole valley on every side as far as you could see was one mass of luxuriant vegetation of indescribable strangeness.

* Edward Peak of Mt. Baker.

FLOWERS ON THE TERRACE OF BUAMBA

From Fort Portal to Bujongolo—Mobuku Valley.

The ground was carpeted with a deep layer of lycopodium and springy moss, and thickly dotted with big clumps of the papery flowers, pink, yellow, and silver white of the helichrysum or everlasting, above which rose the tall columnar stalks of the lobelia, like funereal torches, beside huge branching groups of the monster senecio. The impression produced was beyond

THE HEATH FOREST.

words to describe; the spectacle was too weird, too improbable, too unlike all familiar images, and upon the whole brooded the same grave deathly silence.

Here and there, where the face of the cliff was so steep and smooth that no other plant could take root, were great golden

133

Chapter IV.

patches of moss. In the bottom of the valley the soft, thick, mossy carpet was strewn with violets and forget-me-nots, which startle the European traveller by the unexpected familiarity of their appearance.

The day was fine, and the Duke of the Abruzzi was far too impatient to consent to stop at Buamba, close to the end of the valley, nearly in sight of Bujongolo. They snatched a morsel in haste, and started once more across the flowery plateau in full sight of a graceful waterfall, framed in foliage and flowers, falling from a steep point on the right side of the valley.

The way proceeded for a certain distance upon the left side of the Mobuku, and then crossed again to the right at the foot of the last rise. The valley is full of traces of the former passage of glaciers, the rocks are worn smooth and streaked; there are moraine piles, boulders, etc., etc.

One last climb up a steep slope some 600 feet high, over mud and stone, brings the expedition to the right side of the valley, where a heap of blocks, surrounded by tree heaths, are overhung by a high rock which forms a shelter. This is Bujongolo, a veritable eyrie, at a height of 12,461 feet, and 2,528 feet above Kichuchu.

The Prince and his companions reached this point about two in the afternoon, leaving the caravan of porters far behind. Most of the latter had stopped at the Buamba shelter, and only a few with a small number of parcels rejoined the expedition that evening.

The place was rough and wild. A cold and biting wind blew off the glacier, and suggested surroundings very different from those usually associated with Equatorial Africa.

The members of the expedition were full of excitement and

134

WATERFALL AT BUAMBA.

From Fort Portal to Bujongolo—Mobuku Valley.

satisfaction at having at last reached the foot of the mountains which they were to explore. The journey from Italy to this point had occupied 54 days.

The first night was spent in the open. The tents had not arrived, and many were without even a sleeping bag. A few sheep had come so far with the porters, and frightened by the strange place huddled around them. The shapes of the naked blacks crouching around a great fire showed dimly in the night.

Cagni, barely convalescent, had left Entebbe two days before and was hastening, by forced marches, to join the expedition and take part in their work.

CHAPTER V.

PEAKS AT THE HEAD OF MOBUKU VALLEY.

Organization of the Base Camp at Bujongolo—Upper Mobuku Valley—
H.R.H. starts on the first exploring party—The Mobuku Glacier—Camp on
the edge of the Glacier—Terminal Ridge and Grauer Rock—First comprehensive
view of the Ruwenzori Chain—First Ascent of the Kiyanja Peaks—Vittorio
Sella at Camp I—Photographic work at Grauer Rock—Fog, Snow and
Storms—Sella ascends a third peak of the group—Difficult descent to
Bujongolo—Four days of bad weather—Camp Life—A leopard visits the
camp—The journey of Commander Cagni from Entebbe to Bujongolo.

ON the morning of the 8th June the
Bakonjo porters who had stopped the
day before at Buamba arrived at
Bujongolo in small detachments.
Meantime the Duke and his companions, feeling very stiff and un-
comfortable after a night spent upon
the rocks, in the open air, deliberated
as to the best means of forming a
camp under the existing conditions.
At first sight the thing seemed
impossible. Great blocks of rock, heaped in confusion at
the foot of the cliff, or projecting from the hollow at its
base, left not a yard of ground free. Underneath; the pile
of blocks formed caverns and hollows of which a few were

138

relatively dry and big enough to form possible shelters for the natives.

The cliff overhung the place, while the chaos of loose blocks reached to the verge of the steep slope which led down to the bottom of the valley. This slope was one tangled mass of moss, mud and stones, shaded by the desolate heath forest.

BUJONGOLO.

They began by cutting down numerous trees, and so distributing the trunks among the rocks as to build up platforms wide enough to carry the six tents. These tents stood on different levels, making two groups separated from one

Chapter V.

another by a huge boulder. To pass from one of these groups to the other you had either to go round the boulder, under the perpetual drip of water which, even in fine weather, came off the edge of the overhanging rock, or else you must clamber between the boulder and the rocky wall, a feat requiring some acrobatic skill. Close to the tents, in a small space between three heath trees, were arranged the instruments which composed the small meteorological observatory.

By no effort was it possible so to transform this inconvenient spot as to create an even tolerable camp, such as would have been desirable for a base station, where the expedition might spend a considerable time, and whither exploring parties might return from the high mountains for rest and refreshment. Unfortunately, there seemed to be no place in the region which combined other attractions with a certain measure of shelter from the weather.

The River Mobuku flows at the foot of Bujongolo more than 600 feet below. The camp stood nearly at the entrance of a little tributary valley, which at this point opens out of the right flank of the Mobuku Valley. The latter is visible for a short tract only, not farther than the foot of Kiyanja, at which point it makes a sharp bend to the northward. Kiyanja has from this side the appearance of a high rocky wall ending in a sharp peak. To the left of this peak, at the top of the wall, lies a level glacier overhung by a rounded summit. To the right stretches a jagged ridge, at whose feet flows down into the valley another glacier, partly hidden by the corner formed on the left slope of the Mobuku Valley at the point where it turns to the north.

Opposite the camp, on the other side of the valley, a spur descends gradually down to the plain of Buamba. Beyond this

BAKONJO PORTERS

spur rises a great double peak.* Two ridges run up straight from the base to the points forming a wide couloir between them.

At this elevation, where the temperature often reached freezing point during the night, it became indispensable to

THE KIYANJA OF JOHNSTON FROM BUJONGOLO.

clothe the Bakonjo porters in some degree. The Duke had foreseen this, and warm flannels and blankets were distributed among them. They had great difficulty in putting them on, and their long and ludicrous attempts generally resulted in a frantic effort to squeeze their legs into the sleeves of the

* Mt. Cagni.

141

Chapter V.

woollen vests. The blankets tied around their shoulders and girt with a rope around the waist formed a garment somewhere between a toga and a cassock. At all events, the poor fellows were now protected from the cold, which was the essential point.

While the Duke, with the help of Dr. Cavalli, directed the organization of the camp, Messrs. Knowles, Sella and Roccati made a preliminary exploration as far as the Mobuku Glacier at the head of the valley.

On the following morning, June 9th, Mr. Knowles and Mr. Haldane, who had accompanied the expedition as far as the foot of the mountains, and used all their authority and their great experience to facilitate its progress, left it definitely and returned to Fort Portal. H.R.H. remembers with gratitude the invaluable help which they gave to his enterprise.

The porters went down to fetch the loads which had been left behind at Kichuchu. The Duke, with his guides and Botta and five Bakonjos, started for the upper end of the valley.

After leaving Bujongolo, the way continues to skirt the right slope of the valley. The bottom of the valley is nearly level, marshy, dotted with reeds, lobelias and senecio, and strewn with fallen trunks upon which you stumble at every step, and slippery with wet mosses in which you sink to the knee. The opposite side of the valley consists of a smooth rock wall.

Where the valley turns northward it grows still narrower, forming a gorge between steep walls. At the upper end the Mobuku Glacier appears actually to overhang it, all broken and full of crevasses, covering the upper portion of the last rocky cliff and ending in an ice cavern whence issues the

UPPER GORGE OF THE MOBUKU VALLEY.

torrent. Near to the glacier the only plants are arboraceous
senecio, several yards high.

A little before the end of the valley the way crosses the
torrent and mounts a frontal moraine left by the retreating
glacier. In this way a projecting rock is reached where
Grauer had encamped at a height of 13,229 feet above the
sea, a little below the lower end of the glacier. This was the
last point where it was possible to light a fire, and during
the brief halt the porters huddled shivering around the flame.
The distance from Bujongolo to this point is one hour's march.
The way now continued skirting the rock to the right and
ascending a short chimney closed at the top by a boulder, from
which still hung one of the ropes placed there by Grauer to
facilitate the ascent. Thanks to this assistance the obstacle
was easily surmounted.

In one more hour's climb up the rocks the left margin of
the glacier was reached just above the terminal fall of séracs.
Here the way skirted the glacier for a bit, and then proceeded
again to climb the rock wall over a difficult passage, which
the porters would have been unable to negotiate without the
assistance of the guides. Their bare feet slipped continually
upon the smooth steep moss-grown slabs of rock, or got
wounded on their edges and sharp points. At last the Duke
gave up the idea of bringing them further and sent them
back to Bujongolo.

A short traverse led back to the glacier at the foot of a
rocky projection. The Duke had wished to encamp on the top
of the ridge so as to be there at the following dawn when
there was greater chance of a clear sky. But hardly had
they reached the glacier before a dense fog enveloped the
party and shut out everything from their sight. It was

145

Chapter V.

impossible to get on any further on that day. With their ice-axes they levelled a little space between the stones and here set up the one Whymper tent which they had brought with them.

After Bujongolo there were no more names for any of the places, and therefore the subsequent camps are indicated by numbers. This one on the rocks to the left of the Mobuku Glacier, above the terminal ice-fall, was Camp I, altitude 14,118 feet. Botta and Laurent Petigax at once redescended to Bujongolo. Joseph Petigax, Ollier and the porter Brocherel remained with H.R.H. The afternoon passed slowly and tediously in the cold, damp fog, which did not lift until late in the evening.

Before daylight on the 10th of June, the weather being clear, the Duke, seized by an irresistible impatience to proceed, and dreading a return of the fog at any moment, hurried on the guides at a forced pace down the rocks, on to the glacier, and up the snow slopes with their few crevasses, and in about half an hour reached the top of the ridge. The daybreak had hardly commenced.

The whole range of mountains stood before them, with only the topmost peaks shrouded in mist. They had reached the lowest point of the ridge at the top of the Mobuku Glacier. Here a small peak projected from the snow, covered with black lichens and mosses, while a few grasses and a species of thistle blossomed on its sides. This is the rock which Grauer, in January of the same year, had named King Edward Peak, 14,813 feet.

From this depression, which may be described as a col, the ridge rises to the eastward, on the right, as far as two rocky peaks* separated by a small glacier. Wollaston, with Woosnam,

---

* Moore and Wollaston Peaks.

UPPER END OF THE MOBUKU VALLEY

Peaks at the Head of Mobuku Valley.

had ascended the easternmost of these in February, and had then supposed it to be the Duwoni of Johnston. In the opposite direction the ridge stretches west and south, forming two more peaks, evidently higher than those to the east of the col. These south-western peaks form the Kiyanja of Johnston.*

Moore Peak.　　　　Wollaston Peak.

EASTERN PEAKS OF MT. BAKER, SEEN FROM EDWARD PEAK.

In fact, the peaks at the upper end of the Mobuku Valley form a single group, ending in a continuous ridge, which curves southward in a complete semicircle, circumscribing a vast amphitheatre, covered to a great extent by glaciers.

---

* Semper and Edward Peaks.

147

Chapter V.

To the north, on the other hand, the groups tower above a vast valley where the clear waters of a peaceful lakelet reflect the rocks and glaciers round about. This turned out to be the upper end of that valley which the Prince had detected at its opening into the Mobuku Valley between the two southernmost Portal peaks opposite Nakitawa. As he had then surmised, this valley does actually penetrate to the heart of the range, and is entirely surrounded by snow peaks and glaciers. To the south of it lies the eastern end of Kiyanja, while to the west stands the great central group,* formed, as had been seen from the Hima Valley, of four distinct peaks standing two and two at either end of a ridge whence a great glacier flows down and covers the entire slope. To the north stands the Duwoni† of Johnston, which from this point appears in fore-shortening with two squat snow peaks. There could now remain no doubt but that the two northern peaks of the central group were the highest of the whole chain.

Further off, to the right of Duwoni, behind a great spur which runs down from Duwoni eastward, appeared two more snow peaks‡ standing at the head of a tributary of the Bujuku Valley. On the last ridge of this eastern spur of Duwoni there is a strange monolith, standing up straight as a tower, and with regular angles, which, at a distance, looks almost architectural.

The discovery of the Bujuku Valley proved quite clearly that the terminal ridge at the head of the Mobuku Valley is not a portion of the watershed of the chain, as had been supposed by all the Duke's predecessors who had come so far. It furthermore proved that neither the main group, including the

---

* Mt. Stanley.   † Mt. Speke.
‡ Mts. Emin and Gessi.

Mt. Stanley.   Edward Peak.   Mt. Baker.   Moore and Wollaston Peaks.

MT. STANLEY AND MT. BAKER, TAKEN FROM THE STAIRS PEAK OF MT. LUIGI DI SAVOIA.

highest summits, nor the Duwoni of Johnston, has any connection with the Mobuku Valley. H.R.H. was the first to behold the complete panorama of the range spread out before his eyes. It

THE HIGHEST PEAKS AND LAKE BUJUKU, SEEN FROM GRAUER'S ROCK.

was a far more imposing sight than could have been imagined by those preceding explorers who, once they reached the terminal gorge of the Mobuku Valley, supposed that the

151

Chapter V.

glaciers and peaks around them were the most important of the whole chain. Wollaston alone had had a glimpse of the groups to northward, but the fogs had not permitted him to appreciate their number nor their exact situation. Even in the former attempts to explore the range from the west, single mountains only had been visible. Possibly David had had a wider view, but his description is vague and confused.

It was barely 6.30 in the morning when the little party once more set out towards the west in the direction of the highest peaks of the group, proceeding over hard snow broken by a few crevasses on the left flank of the crest facing the Mobuku Valley.

The ridge rises first to a peak* of broken and rotten rock (15,843 feet), of which H.R.H. reached the summit at 8 a.m. A light wind was blowing from the western valley, and drifts of mist began now and again to shroud the prospect from their sight.

To the west of this peak a jagged and slightly marked *arête* leads precipitously down to the pass which connects Kiyanja with the central and highest group. The main ridge, on the other hand, bends southward, and from thence onward forms part of the watershed of the range between the Mobuku to the east and another smaller valley which falls away westward toward the Semliki. The west face of Kiyanja is precipitous like the north face, which overhangs the Bujuku Valley.

Without stopping on this first peak, the caravan proceeded southward along the ridge towards the highest point, now less than 400 yards distant. At 9.15 a.m. the Prince was the first to set foot upon the highest summit of Kiyanja,† 15,988 feet. The rocks of the summit are covered with fulgurites in the form of

---

* Semper Peak.   † Edward Peak.

vitreous efflorescences. The wind had ceased, and everything around them was enveloped in mist. The temperature was mild, about 43° F. They remained four hours on the top, looking anxiously for any opening in the mist in the hope of gathering further details of the scene around them.

Semper Peak.   Edward Peak.

MT. BAKER (THE KIYANJA OF JOHNSTON) SEEN FROM THE WEST.

They did not wait in vain. They were able to make out that the watershed ridge proceeded southward from the peak upon which they stood and downwards to a col beyond which was another group of mountains,* a short chain of ridges and rocky peaks, with a few glaciers of far less extent than those of

---

* Mt. Luigi di Savoia.

153

the northern groups. The low watershed col had every appearance of forming an easy pass between Bujongolo and the valley
to the west of the Kiyanja, by which it would be possible to
reach the foot of the central group without difficulty.

Before dipping down to the col, the south ridge of the
Kiyanja rises once more into a knob of rock, which is clearly
visible from Bujongolo, and is the point ascended by
Mr. Wollaston in February and in April.

By one o'clock the party started back. They crossed once
more the peak which they had ascended first, and proceeded in
a fog, which was now dense and immovable, over the tracks
which they had left in the snow in the morning. At three
o'clock they reached Grauer's rock, and in half an hour more
were in the camp beside the Mobuku Glacier. Here they
found Sella, who had come up with Laurent Petigax and Botta.
With the help of six natives they had brought up a second
tent and the photographic apparatus. Sleety rain was now
falling, which soon turned into a thick fall of snow.

On the morning of the 11th, the Duke returned to
Bujongolo. Sella, with Botta and Brocherel, in their turn,
ascended the col. The tracks of the Duke's party had
disappeared under the new snow which had fallen during the
night, and the mist made it impossible to see even a few
steps ahead.

After a few hours spent on the col in vainly waiting for
the mist to clear, they proceeded to climb the rocky crag which
Grauer had named after King Edward. Once on the top, in
spite of the snow which was now falling again, they set up the
photographic camera on its tripod, and, huddling around it,
waited patiently. By two in the afternoon Sella gave it up,
folded up the camera, and was on the point of leaving the

peak, when suddenly the sky began to clear. The mists melted rapidly on every side, and in a few minutes all the mountains were uncovered except only the extreme summits. The camera was immediately set up again and a panorama taken.

A clear sunset followed. The sun went down just over the two highest points, lighting up the snow with its last rays. At nightfall the storm began again with thunder, lightning and heavy snow. Sella returned to the col in the morning. He saw the mountains once more, but under a leaden sky with diffused light and no shadows. Dark banks of mist were drifting upward from the east in a light wind, and settling little by little over valleys and peaks.

From the col Sella proceeded to a peak to the east,[*] 15,269 feet, over rocks which were not difficult, but here and there were rendered dangerous by the ice and snow.

The snowfall had again begun, but he remembered his luck of the preceding day, and waited patiently on the top until three in the afternoon, but without success. On returning to the col, he found Roccati, who had come up from Bujongolo with a guide to take observations on the glaciers. In the evening Sella remained alone in camp with Botta. The snow was now falling thick and continuous, without the smallest interval.

On the following day, 13th of June, the tent was folded up to return to Bujongolo, whence five natives had come to fetch the loads. The descent was far from easy. Numberless rivulets, now swollen with water, fell in little cascades across the narrow muddy path, and made the mossy rocks even more slippery than they had been. It was extremely difficult to induce the terrified natives to proceed. In the

---

[*] Moore Peak.

Chapter V.

chimney, near Grauer's camp, which is overhung by a projecting rock, they had to pass under a real waterfall, with a certain risk of being carried away down the precipitous slope. Here and further down, as far as the point

BUJONGOLO.

where the path becomes less steep, Sella and Botta were obliged to carry all the loads themselves by instalments, while the Bakonjo, silent and passive, could scarcely manage to proceed at all. Finally, about seven o'clock in the evening, drenched with water and covered with mud, they reached their companions at Bujongolo.

Here from the 11th up to the 14th inclusive there was no improvement in the weather. The rain was nearly continuous, while storms of wind, with thunder and lightning, followed

Peaks at the Head of Mobuku Valley.

upon one another at short intervals. Thick dark fog enveloped everything. The camp was soon invaded by mud and water, and a continual drip came down upon the tents from the overhanging rock. Under these conditions it became difficult to kindle a fire, and the only resource was to keep it burning day and night, which required no mean amount of work in feeding it and providing sufficient fuel. On one side of the

HEATH FOREST BELOW BUJONGOLO

huge boulder which divided the camp in two, stood the three tents of the Duke, his companions, and Bulli, standing on different levels. They had constructed, close by, a rough shed where they could eat, and the kitchen was near at

157

Chapter V.

hand. On the other side of the boulder the guides, by dint of displacing large rocks with their ice-axes and working hard at levelling, had made room for their own tents.

Every time that anyone stepped out of the camp he would sink into the mud. It was impossible to circulate between the tents without nailed boots, because the moment that you came out a sort of mountain-climbing gymnastic began, where it was necessary to hold on tight at every step.

The mean temperature was from 39° to 41° F. At night it generally fell to 33°–34°, and sometimes to freezing point. The dampness, however, was far more trying than the cold.

One event alone would occasionally relieve the tedium, namely, the arrival of the post. The letters were brought up by swift couriers—wrapped up carefully in banana leaves, and stuck in the end of a cleft cane.

Now and again the camp would be filled with pungent smoke, extremely irritating to the eyes and chest, which came from the fires lit by the Bakonjo in the underground cavities between the boulders. They huddled together all day long in these dens, where they had not room to stretch themselves out at length, and ate or smoked incessantly when they were not asleep. Their real providence was the fire. They never left it except when called away, and rushed back to squat around it as soon as they were no longer required. They carried it about with them from place to place, using a sort of dry fungus which remains kindled like tinder, and which they keep in a case made of banana leaves. The moment a halt was called during a march, in less than no time the natives would have kindled a fire and be enjoying a fine blaze and smoking their pipes, and it was not always easy to induce

BUJONGOLO

them to start again promptly. They were once found on the road, shivering in the rain and stark naked, having taken off their vests and blankets so as to enjoy to the full the heat of the glowing coals. They ate eagerly whatever food was supplied to them, but they did not like novelty. They made wry faces before making up their minds to swallow tea, and far preferred their mess of dura flour, which to us seems nauseating to the white wheat flour, even when prepared with butter.

In spite of these trying conditions of life, the Bakonjo showed admirable patience and docility. It was very rare, indeed, for even a single porter to refuse to go on with his load, although they nearly always got their feet swollen and hurt by the stones.

Once only, on returning from Bujongolo after a few days of hard service in the mountains, ten of them deserted because it was not possible to dismiss them, as they desired. The law of Uganda does not permit native porters to abandon a caravan conducted by whites until they have finished the time or traversed the distance for which they have engaged themselves. It came out afterwards that the deserters had been instigated by a native convert of the plain, the one and only Baganda who had been willing to follow the caravan as far as the mountains. Notwithstanding the cold and the bad weather, they ran away stark naked, after honestly depositing their warm clothes and blankets near one of the tents. During the sojourn of the expedition at Bujongolo, a certain number of natives got bronchitis and coughs and were sent down. One got his feet frostbitten and was carried as far as Fort Portal to the hospital.

Poor Igini, the cook, had the hardest life of all. He was the only one whose activity was confined within the

159

Chapter V.

ring of deep mud which turned the camp into a close prison. Squatting between four boulders, hedged about by the cases of rations, the kitchen implements, the fire, and the tent, he had far less chance of exercise than during the polar winter which he had spent in Teplitz Bay, where he was forced to go half a mile to fetch the meat of some bear hung up by the ship, or had to work to disinter the cases of rations, or help to run after the dogs.

The daily visits of a big leopard, whose den was situated under the heath forest in the neighbourhood of the camp, did not contribute to render the sojourn at Bujongolo pleasant. This leopard was observed for the first time, on the 11th of June, by a native, not far from the tent devouring two sheep belonging to the expedition. On the following night he prowled around the camp. On the evening of the 12th, the Duke, who was absorbed in writing at the opening of his tent, saw him only a few paces off. The animal fled as soon as he stood up, but his boldness gave cause to fear for the porters who slept unprotected, or for those who went to fetch water for the camp. In vain were the surroundings examined and beaten; the animal seemed very cunning, and when hunted never allowed himself to be seen.

On the evening of the 13th, the weather had shown a tendency to improve, but during the night grew worse again. On the evening of the 14th, however, it seemed really to clear. The sky became free from clouds, the last traces of mist melted away, and all the mountains came into sight covered far down with the fresh snow which had fallen during the last days. The intolerable imprisonment was finally coming to an end, and the Duke made preparations to start out on the following morning to explore the central group of the chain.

MT. CAGNI SEEN FROM BUJONGOLO.

Peaks at the Head of Mobuku Valley.

Commander Cagni, who was hastening up the Mobuku Valley, had by this time nearly rejoined his comrades, who believed him to be still many days' journey off.

He had left Entebbe, as we said, on the 5th of June, with twenty-five porters, a rickshaw and a horse. In a short time he so far recovered his strength and got so perfectly into training that he was able to make two, or even four, stages in a day. He took advantage of the full moon to leave before dawn, and continued the march till late in the day, doing 25 to 27 miles at a time. The porters, allured by presents of a sheep or a little money, performed miracles. Once they marched for seventeen hours, covering 32 miles.

In six days, Cagni reached Toro, where King Kasagama showed him every courtesy. He left again in the morning of the 12th of June. At Butanuka he found the 178 Baganda porters who had been sent back from the Mobuku Valley. Following the directions of the Duke, he dismissed a portion of them, and sent the others back to Fort Portal, there to await the return of the expedition from the mountains. He had difficulty in crossing the Wimi River, which had now become an impetuous torrent some 50 yards wide, with a depth of more than three feet at some points, and he found a still more serious obstacle in the Mobuku, swollen by the same rains which were imprisoning the expedition at Bujongolo.

Not having a rope long enough to permit of his stretching it across the river, as the expedition had done, he tied together the halter of his horse, the tent ropes, the cords used to tie the loads, etc., doubling them several times, and in this way he contrived a rope long enough to cover about half the width of the torrent. This he had kept taut across the central and swifter part of the current by two groups of men. Thanks

Chapter V.

to the willing help of the chiefs and natives of the neighbouring villages, he managed to cross the river without accidents, but with the loss of half a day.

On the 14th of June, at Bihunga, where the native soldiers of the escort were established, he changed his Baganda porters for Bakonjo. Two days later he reached Bujongolo, thus accomplishing the whole journey in ten stages.

Here he found only Dr. Cavalli, the Duke having left the evening before, while Sella and Roccati had started that very morning to ascend the col to the south of Kiyanja.

CHAPTER VI.

Peaks of the Central Group.

The Bakonjo's dread of the Western Slopes—H.R.H. leaves Bujongolo—March through fog and mud—The Col on the Watershed—Camp near the Lake— Ascent of the Valley to the West of Kiyanja—Camp III—Col at the foot of the Central Group — Camp IV — The Bujuku Valley once more in sight—Ascent of Alexandra Peak—In the Fog—Climbing Margherita Peak— Snow-blindness—Elena and Savoia Peaks—The Expedition united again— Adventures of the Duke's companions from the 15th to the 20th of June— Sella and Roccati climb a peak of the Southern Group.

DOUGLAS FRESHFIELD had been told by his caravan leader that the col towards which the southern ridge of Kiyanja runs down, and which forms a part of the watershed, had been used as a pass by the natives on the west of the chain who were in the habit of crossing it into the Mobuku, on their way to Buamba to trade with the Bakonjo.

The Duke, however, failed to gather from his porters the least scrap of information as to any way of communication between the eastern and western slopes. They appeared rather to experience a feeling of terror for the district beyond the ridge, and seemed

165

Chapter VI.

profoundly convinced that to go towards the Congo was equivalent to going to meet certain death. It was only too natural that, with these feelings, they should show extreme reluctance to following the Prince westward.

On the morning of the 15th June, there were only nine Bakonjo at Bujongolo, hardly a sufficient number, with the addition of the four guides and Botta, to carry the kit of the Duke, now reduced to absolute necessities and rations for a few days. At the last moment the natives put forward a claim to being paid every day, and the Duke of the Abruzzi was obliged to load himself with a not inconsiderable weight of rupees.

Finally, about eight o'clock, as no other excuse for fresh delay was available, they started from camp in brilliant sunshine. First they went up the little valley which, as we saw, opens on the right-hand side of the Mobuku, near to Bujongolo. They followed the line of the small torrent, crossing from one bank to the other, and so reached the top of the spur, and came into a valley formed by a torrent fed from the southern glaciers of Kiyanja. This is the same torrent which forms the picturesque waterfall on the right slope of the plain of Buamba. Close to the head of the little valley are two projecting rocks forming natural shelters, similar to those of Kichuchu and Buamba.

The ground was drenched with the rain which had fallen during the preceding days, and after an hour's march everyone was wet to the skin, and covered with mud. The march was tiring, because at every few steps you slipped or sank into the mud. The porters were suspicious of the unknown country towards which their steps were directed, and proceeded unwillingly, with exasperating slowness. They had stopped

THE VALLEY TO THE WEST OF MT. BAKER.

Peaks of the Central Group.

twenty minutes after leaving Bujongolo, and had immediately kindled a fire and lit their pipes. After another half-hour's march they repeated this performance. When urged to proceed they would answer by pointing to their head, feet, belly or legs, each of which portions of their persons seemed to have been suddenly afflicted with painful complaints. To make

MT. STANLEY SEEN FROM FRESHFIELD'S COL.

things worse, the fog closed in again, and the marshy valley was crossed without a glimpse of anything beyond the senecios and lobelias around them, and the moss, mud and stones at their feet. The watershed* was at last reached after an easy climb, partly on the slope and partly in a gorge.

* Freshfield Pass.

Chapter VI.

A cold wind was blowing and the porters rushed to find shelter. The height was 14,193 feet above the sea-level. They were above the zone of trees, and there were only mosses, lichens and clumps of everlasting flowers.

The wind drove the fog hither and thither, opening up glimpses of the country now in one direction, now in another. To the north of the col rose the southern ridge of Kiyanja, wide, rounded, and covered on the top by a glacier which falls over to the right and left on the two slopes, and which must have once come down so as to cover the entire col. The traces are clear on the polished and lined rocks. Southward stands the group of rocky peaks which H.R.H. had already observed from the summit of Kiyanja. Here they saw two small glaciers which fill two cols, while a third between them is rocky and free from ice. Four peaks form these cols; the westernmost and furthest off appeared to be the highest.

At the foot of these peaks, between them and a spur of Kiyanja, lies a valley which slopes down due west. Beyond this spur they caught sight of the light reflected on two tiny lakes, which lie at the bottom of another valley running from north to south, starting from the col between Kiyanja and the central group.

It was this col which the Duke wished to reach and to make his base for attaining to the highest peaks. While he carefully noted down every detail of the country which was visible, taking advantage of the rifts in the moving mists, a guide went forward to find out whether it might not be possible to skirt the western slopes of Kiyanja without descending to the bottom of the valley, which would then necessitate reascending to the farther col. The guide now came back and brought news that it would not be possible to skirt the mountain because its side

LAKE TO THE WEST OF MT. BAKER—CHARRED SENECIOS.

VALLEY TO THE WEST OF MT. BAKER

towards the valley was precipitous. They were therefore obliged to go down to the lakes.

Soon after midday the little party, leaving a portion of its loads on the col so as to move more rapidly, started afresh, and first skirted the western slope at the same level, very little under the pass but above the forest of senecios, in order to reach the ridge of the south-west spur of Kiyanja.

From here they descended towards the lower of the two little lakes. The descent was steep, the mud was slippery, and their way led through a forest of senecios and clumps of helichrysum, which the guides cut and broke with blows of their ice-axes to clear a path. There were great smooth slabs to be avoided, which here and there stuck out of the ground and were too steep to walk upon. The porters kept striking their loads against the low and dense ramifications of the senecios, slipped on the big stones, stumbled among the dead trunks and branches which lay half-buried in the mud, and had to be incessantly encouraged and urged to proceed. They were overhung by the precipitous sides of Kiyanja, which threatened them with stone falls. As they neared the bottom of the valley, they were surprised to find a vast tract of senecio forest, where the trunks and branches were bare, blackened, and partially carbonized by recent fire. There was no sign to indicate the passage of man, nor was it probable that the inhabitants of the valleys would have pushed so far up without cause; hence it must be supposed that the fire was either spontaneous or caused by lightning. The dense mantle of dead leaves which hangs downwards around every branch of the senecio under the terminal bunch of green leaves, and which is one of the chief features in the strange aspect of this curious plant, offers abundant fuel for fire and is as easy to

171

Chapter VI.

kindle as tinder. It might perfectly well be set ablaze merely through the heat developed in fermentation, which must be especially active and energetic in this climate. The fact is interesting were it only to show that there must be occasional dry spells of sufficient length to allow of the dead leaves getting dry to a certain extent, for in the soaked state in which they are usually found, it would seem quite impossible that they should provoke a conflagration. Between the blackened skeletons, striplings were already shooting up to replace the ruined forest.

The party reached the shore of the first lake towards four o'clock. It was plainly impossible to oblige the porters to proceed on that day. The Duke decided to encamp upon a spur which runs out into the lake, and is about 90 feet high. The level above the sea was 13,271 feet; 810 feet above the camp of Bujongolo.

The valley, which up to this point ran due north and south, here turns abruptly to the west, becoming so narrow that the lakelet fills the whole bottom, which forms a sort of oval basin resembling a crater at first sight. A few ducks wander over the water. There were traces of leopards and marmots, and a few crows were flying overhead. There was no other sign of animal life. A little torrent, falling down from the glaciers of Kiyanja, gave abundance of water. There was plenty of wood within reach, and it was soon possible for everyone to dry his garments around a big fire.

In the clear fine afternoon the little lake, barely ruffled by the breeze, reflected the snowy peaks. The valley was one mass of flowers, and in this peaceful scene they soon forgot the fatigue of the hard day. The sun went down behind a dense layer of clouds, which lay across the sky to the westward.

THE UPPER LAKE IN THE VALLEY TO THE WEST OF MT. BAKER.

Peaks of the Central Group.

Presently the sun appeared again below these clouds and lit up the western sky, the valley, and the vast forest of the Congo, which merged in the intense red of the far-off horizon.

On the following morning, Laurent Petigax, Brocherel, and three native porters went back to fetch the loads which had been left on the preceding day on the pass when they came up from Bujongolo. The others set out on the march carrying the rest. They skirted the two little lakes at the foot of the slopes of Kiyanja, cutting their way step by step through the dense tangle of senecio and helichrysum. Between the thickets of sempervivum were found specimens of an exceedingly beautiful large flowering hypericum, together with ranunculus, several plants of the cruciferous family, alchemilla, balsam, robbia, etc. The valley was narrow and grim, shut in between steep rocks, dryer than the Mobuku, and showing many and clear traces of glacier action at a not very remote period. The whole bottom was full of moraine rubbish, mingled with detritus, fallen from Kiyanja. Both lakes are of glacial formation. Under the lower lake lies a frontal moraine forming a dam, with a spur of rounded rock. The two lakes are divided by a rocky transversal ridge covered with detritus. To the north of the upper lake lies another moraine forming a steep bank, above which a high plain slopes upward. Here the valley widens out about two-thirds of a mile. At the foot of Kiyanja is a long névé, formed by avalanches falling over the side.

The valley is dominated by the southern peaks of the central group, from which flow down two glaciers, forming two sources of the torrent; the third springs from a glacier of Kiyanja.

Chapter VI.

Camp III was formed at a height of 13,842 feet, almost directly under the two peaks of Kiyanja, which the Duke had ascended six days before. Here also was an abundance of water and of fire-wood.

HEAD OF THE VALLEY TO THE WEST OF MT. BAKER.

From this point the ascent of the col* offered no difficulty. The way followed the ridge of an ancient

---

* Scott Elliot Pass.

Alexandra Peak   *Margherita Peak*

THE HIGHEST PEAKS

median moraine, which shows that at one time the glaciers of Kiyanja joined those of the central group and ran down into the valley together. The senecios and everlasting flowers came nearly up to the top of the col, growing gradually less dense.

On reaching the top of the col, the party followed the ridge westward as far as a point near the edge of the glacier, which runs down to the feet of the southernmost peaks of the central group. These peaks form two imposing

CAMP IV, CLOSE TO THE ELENA GLACIER.

towers of rock. The camp was pitched upon the broken stones near the glacier at a height of 14,817 feet.

There were now rations for one day only, so the whole of the Bakonjo, as well as Laurent Petigax and Botta, were sent back to Bujongolo. Joseph Petigax, Ollier and Brocherel remained with the Duke.

The afternoon was clear and the mountains free from cloud. The camp overlooked the valley with its little blue lake, which had been seen from the ridge at the top of the

Chapter VI.

Mobuku Glacier. From the col which they had just traversed, a nearly perpendicular rock wall falls down to this valley on the north. The eye followed the valley for a long distance eastward and saw it turn southward in the distance to join the Mobuku. Thus there remained no possible doubt as to this being really the Bujuku Valley, and as to the great snowy mountain to the north being really the Duwoni of Johnston. To the south-east the view is shut out by the mass of Kiyanja.

The guides who had climbed the glacier to make out the way to the central group returned towards evening. The sunset was less clear than on the preceding days. The extreme nearness of the goal made the forced delay intolerable. The Duke, cooped up with the guides in the narrow space of a single tent, passed a great portion of the night in anxious watching, preoccupied by disagreeable doubts as to the weather.

Finally, the day dawned on the 18th of June with a clouded grey sky. They roped together hastily and in silence. Joseph Petigax and Ollier came first, then the Duke, and lastly, Brocherel. They began the ascent of the glacier along the way traced by the guides on the preceding day. The great ice plain was reached without difficulty in about one hour. It was 6.30 in the morning, and the peaks which they desired to reach stood before them at a very short distance. They were both covered with snow, and the southernmost, which stood nearest to them, showed a rock precipice on the east side surmounted by a big cornice of snow and was joined by a rounded ice col to the northern peak, which was somewhat higher, and from which ran down two ridges, one eastward in a straight line towards the valley, the other north-

178

westward, slightly concave, and terminating in a characteristic shoulder. The peak and the ridges are overhung by a gigantic cornice, supported by a colonnade of icicles and aiguilles of ice which at a distance seem like a fine white lacework.

Round about them the snow had the pale, lack-lustre hue of bad weather. For one moment a ray of sunlight lit it up, but was immediately quenched by the dense clouds which were piling up rapidly from the east. Gusts of wind were

ALEXANDRA AND MARGHERITA PEAKS FROM THE STANLEY PLATEAU.

blowing from the east, and layers of mist were ascending the valley in compact masses and soon shut in the party completely. They continued their way in silence. Without the least uncertainty, with a sure intuition of the right direction, Joseph Petigax made his way through the dense mists and ascended the high plain as far as the foot of the south-east ridge of the southernmost peak. The hard snow, which carried them, allowed them to climb swiftly up the wide snowy ridge, cutting only a few steps at the steepest points. By 7.30 a.m. they reached the top of the first peak.

CLIMBING THE SOUTH-EAST RIDGE OF ALEXANDRA PEAK.

A strong wind was blowing from the east. Round about them the dazzling white of the mist was impenetrable to the sight. Everyone had his mind fixed upon the highest summit, which was only a few hundred yards off, but quite invisible, and they stood there waiting and turning their eyes obstinately northward. During an hour and a-half

180

ALEXANDRA PEAK FROM THE SOUTH SHOWING THE SOUTH-EAST RIDGE

there were only brief moments when the mist would grow slightly thinner, so that they could just make out the uncertain outline of the highest peak.

There were only two ways to reach it. They must either descend to the gap and thence attempt to scale the ice wall overhung by the formidable cornice, or they must return to the plateau, cross it under the col, and ascend by the east ridge, a long and indirect way, which would have to be done in the fog without any sign to guide them. As to the question of giving up the ascent for that day and returning to camp, a glance at the set determined faces of the guides was sufficient to show that such an idea never entered their minds.

By nine o'clock they could not endure waiting any longer,

CLIMBING ALEXANDRA PEAK.

and they decided to take the mountain by the shortest route, which was also the more' dangerous, and one after another they started down the slope which led to the col. They

Chapter VI.

proceeded with their faces turned to the wall, placing their feet with caution in the wide steps which Petigax cut in the snow, which was fortunately hard and bore them well.

The col is a narrow strip of ice between two wide crevasses (*bergschrund*); these crevasses pass from one peak to another without a single bridge. It was impossible to turn to the right or to the left; they could only go straight forward to the ice wall, which they could barely make out through the mist.

Where the slope commenced to become steep they put down their rucksacks and other unnecessary impedimenta, and Petigax set to work again. They soon stood nearly vertically one above the other, climbing slowly by the steep steps which Petigax cut in the ice wall, showering down a hail of snow and ice upon the others. Below them the wall was almost immediately swallowed up in the mist, so that they seemed suspended over a bottomless abyss.

In this way they reached the bottom of the cornice where the pendant icicles, joining the upright ice needles, formed a colonnade as thick as the trees in a forest, upon which rested the heavy snow-dome whose solidity was open to doubt. The effect seen through the mist was strange and weird. In their insecure position, holding fast to the steep slope, they had to climb around the ice columns to reach the point where the cornice jutted out from the ice wall in order to find a passage. This passage they found in a cleft of the cornice which formed a narrow vertical gully some six feet high. Ollier, standing firmly upon a wide step, served as a ladder for Petigax, who climbed on his shoulders and then upon his head, with his heavy nailed boots, and stuck his ice-axe firmly in the snow above the cornice. In this way he hoisted himself on to the top. It was easy enough for the others to join him. The ridge was now

182

ELENA AND SAVOIA PEAKS, AS SEEN FROM THE STANLEY GLACIER.

MARGHERITA PEAK FROM ALEXANDRA PEAK

vanquished. In a few minutes H.R.H. set foot upon the highest peak of Ruwenzori.

They emerged from the mist into splendid clear sunlight. At their feet lay a sea of fog. An impenetrable layer of light ashy-white cloud-drifts, stretching as far as the eye could reach, was drifting rapidly north-westward. From the immense moving surface emerged two fixed points, two pure white peaks sparkling in the sun with their myriad snow crystals. These were the two extreme summits of the highest peaks. The Duke of the Abruzzi named these summits Margherita and Alexandra " in order that, under the auspices of these two royal ladies, the memory of the two nations may be handed down to posterity—of Italy, whose name was the first to resound on these snows in a shout of victory, and of England, which in its marvellous colonial expansion carries civilization to the slopes of these remote mountains."*

It was a thrilling moment when the little tricolour flag, given by H.M. Queen Margherita of Savoy, unfurled to the wind and sun the embroidered letters of its inspiring motto " *Ardisci e Spera* " (Dare and Hope).

The wind was blowing up rather fresh from the south-east with a temperature of 23°·4 F. Calculations from the observations taken gave a height of 16,815 feet for Margherita Peak, and 16,749 feet for Alexandra Peak. It was now 11.30 a.m. They had taken about half an hour to get down from the first peak to the col, and an hour and a-half to climb from the col to Peak Margherita. These hours were full of intense excitement, owing to their perpetual fear of seeing the way blocked by some insuperable obstacle.

Margherita Peak is all covered with snow, and not a single

---

* *See* " Geographical Journal," February, 1907, p. 138.

rock comes to the surface. The eastern and western ridges seemed to offer easy routes to the summit.

They remained less than half an hour on the peak. There was no hope of the mists disappearing that day, and after finishing the barometric and thermometric observations, and enjoying the first enthusiasm of victory, they began to feel the penetrating cold of the wind. There was an impressive sense of solitude in perching upon this narrow snowy ridge, with the whole earth cut off from them by the mist. Glaciers, precipices and peaks, valleys and plains, lakes and forests, were all veiled by the dense layer of fog, interposed like a barrier between the burning regions of Equatorial Africa and the eternal Alpine snows.

They re-descended the ice wall, resumed their loads, and returned to Alexandra Peak. By 2.20 p.m. they returned to their tent. A few hours later they were all four stricken with snow-blindness. They had been exposed during the whole day to the dazzling whiteness of the fog, and unable to make use of their black spectacles, with which it was impossible to see anything at all. They spent the night and the following day in the tent, bathing their swollen and weeping eyes with tea.

On the following day, 20th of June, they were all much better, so early in the morning they started from the tent in very fine weather, and returned to Alexandra Peak by the same path which they had taken two days before. The Duke arrived on the top about 7.30 a.m., and worked for a long time at measuring the angles of the peaks and the salient points of the chain. He set out again at 9.0 a.m. Drifts of fog were now beginning to invade the scene. They returned to the high glacier-plain and set forth for the two fine rock and ice peaks which stood at its southern extremity.

Peaks of the Central Group.

Half an hour later they attacked the nearer of the two, starting up a gully on the eastern side. About half-way up, where it was steepest, they left the snow to climb on to the

ELENA AND SAVOIA PEAKS FROM THE RIDGE ABOVE CAMP IV.

rocks to the left of the gully, which were steep and not easy, with few hand-holds very inconveniently disposed. Then they came back into the gully, and followed it up to the top, where there was an indentation in the ridge. Through this they traversed the west side, facing the Congo, and climbing up easier rocks, reached the summit. Here they spent an hour in contemplating the peaks and glaciers which kept appearing and disappearing in the continual formation and dissolution of the mists.

Towards twelve o'clock they once more set out, following the ridge southward. In the narrow indentation between the

CAMP NEAR SCOTT ELLIOT'S PASS.

two points there is a sharp tooth of rock with a precipice which falls down towards the Bujuku Valley. They skirted this easily over the snow slopes on the Congo side. From this point, first over a snowy ridge and then over rocks, they reached the ice-cap which terminates the second peak.

The Duke of the Abruzzi gave the name Elena to the first of these two peaks, 16,388 feet, as a homage to our gracious Queen, and the name of Savoia to the second, 16,339 feet. All the four principal peaks of the central group had now been ascended. They could see the camp which they had left that very morning almost vertically under their feet. They could even hear the shouts of the rest of the expedition, who had arrived at the camp from Bujongolo a few hours before.

Towards 2.30 p.m. they began the descent, not returning on their steps, but proceeding southward along the glacier which covers the peak, and then descending the eastern rock wall as far as a wide gully, which brought them back to the glacier a little above the camp.

A few minutes later the Duke was met with great rejoicing by his companions, and the whole expedition was now once more united.

In the space of ten days H.R.H. had ascended Kiyanja and the four ice peaks of the principal group of mountains, had accomplished an extensive triangulation, and identified the position and distribution of the several peaks in relation to the chief valleys.

The 21st of June was given up to rest. The camp was adorned by clothes drying in the sun on the taut cords of the tents, which were now four in number. The guides spent the day in sleep. The afternoon was misty ; higher up it was snowing. The silence of the mountains reigned around, broken now and again by the roaring of an ice avalanche falling into the Bujuku Valley. The different members of the expedition reported what they had done during the last days.

Chapter VI.

On the 15th of June, soon after the Duke's departure, a party of porters laden with provisions, including baskets of fowls and a few sheep, had arrived at Bujongolo, and the silent and nearly deserted camp had become suddenly full of noise.

With their help, on the 16th Sella and Roccati set out in

Stairs Peak.

MT. LUIGI DI SAVOIA SEEN FROM THE SOUTH RIDGE OF EDWARD PEAK.*

their turn for the watershed col, bringing mountain camp equipment and photographic apparatus. They pitched their tents immediately beyond the col on a sloping rock, and made a shelter of tent canvas for the porters. The fog was dense, the wind and cold biting.

---

* For note, *see* following page.

190

Peaks of the Central Group.

On the following day, during some brief openings in the fog, Sella photographed views from the col and from a rock-point on the ridge near the camp. The very same evening the Bakonjos, who had been sent back by the Duke from Camp IV, arrived with Laurent Petigax and Botta.

Sella Peak.   Weismann Peak.

MT. LUIGI DI SAVOIA SEEN FROM THE SOUTH RIDGE OF EDWARD PEAK.*

Everything was frozen and covered with hoar frost. The natives passed the night around a big fire, wrapped in blankets and cloaks which Sella and Roccati gave up to them. The cold was intense, and a thunderstorm was raging. On the following day, June 18th, Cagni, who, as we have already said,

* The above illustrations complete one another and form a panorama,

191

Chapter VI.

had arrived at Bujongolo on the 16th, and Dr. Cavalli joined their comrades on the col.

On the 19th, Cagni and Cavalli with their caravan of porters descended toward the little lakes to the west of Kiyanja. The morning was clear, and the view was open to the westward as far as beyond the Semliki valley over the forests of the Congo. Sella and Roccati, accordingly, turned their steps to the mountains to the south of the pass with the photographic apparatus.

We have already said that this is an important group of mountains with various distinct peaks, between which are small glaciers. The weather, however, spoiled rapidly, and drifts of vapour, driven by the wind from the east, enveloped the higher ridges. They took advantage of brief clearings in the mist to climb the north-east point of the group, which they reached after three hours of ascent, partly over rocks and partly upon the easy glacier to the west. They remained upon this peak until four in the afternoon, and were several times beguiled by apparent meltings of the fog into setting up the photographic apparatus. Night surprised them on their way down into the valley to rejoin their comrades. They lost the track and wandered about upon the steep slope in the forest of lobelia and senecio, among the dense helichrysum, seeking for a way in the darkness intensified by the mist, and slipping at every step in the mud and on the damp and mossy rocks. At last their comrades heard their shouts and sent two guides to meet them with a lantern. They soon reached the camp near the lake, surrounded by the fires of the natives.

On the following day they all reached the fourth camp, from which they were able to descry the Duke's party on the sky line on the summit of Savoia Peak.

CHAPTER VII.

FORMATION AND GENERAL FEATURES OF THE RUWENZORI RANGE.

Ruwenzori and the "Albertine Depression"—Relation to the Nile Basin—Nomenclature — H.R.H.'s Map — The Six Groups of Snow Peaks — The Watershed—The Distribution of the Valleys—What preceding Explorers saw of the Mountains—Confusion of Names and Topography—Altimetric Measurements—The Geology of the Range—The Glaciers—Flora and Fauna.

THE story of the exploration of the two chief groups of peaks has put us in possession of sufficient data to permit of our giving at this point a systematic description of the Ruwenzori range. A more detailed knowledge of the distribution of the groups of the range and of their position with respect to the valleys will help to render the account of the subsequent work of the expedition both quicker and easier.

The Ruwenzori chain is distinguished by extremely remarkable orohydrographical features. Most of the continents slope gradually from the summits of their mountain

Chapter VII.

ranges down to their high central plateaux and thence to the coast level. In Africa, on the contrary, Ruwenzori rises from the so-called " Albertine Depression," a low district forming a region about 600 to 700 feet below the average level of Uganda, and containing the basins of Lake Albert and of Lake Albert Edward with its northern prolongation, Lake Dweru or Ruisamba.

The whole of this depression forms simply a portion of the western " rift." The " rifts " consist of two gigantic trenches, from 20 to 50 miles in width, running nearly parallel to one another, with an interval of 6° longitude, and cutting through the continent from Lake Nyassa north-ward. The easternmost of the two follows the 36th meridian as far as Lake Rudolph, beyond which it inclines towards the Red Sea. The western rift runs between the 29th and 30th meridian and comes to an end near Gondokoro in the Upper Nile Valley. Either rift includes a nearly continuous chain of lakes and numerous mountains and volcanic cones and craters. Either rift is divided by a transversal watershed into two separate hydrographic systems, one to the north, the other to the south. In the case of the eastern rift this ridge is near Lake Naiwasha, about where the Uganda Railway traverses the depression. In the western rift the watershed is formed by a veritable range of volcanic mountains of which some are still active at the present time. This range divides the chain of lakes into two distinct systems. The southern system includes Lakes Kivu and Tanganika; the northern system, Lakes Albert Edward and Albert.

At the southern extremity of the Ruwenzori chain the rift bifurcates : one branch runs to the east of the chain and terminates at the foot of the heights which enclose the

Formation and General Features of Ruwenzori.

basin of Lake Ruisamba to the north and upon which are situated Toro and Fort Portal; the other branch passes to the west of the range and forms the Semliki Valley and the basin of Lake Albert, and is prolonged for several hundred miles by the upper valley of the Nile. Ruwenzori is thus nearly completely surrounded by the " Albertine Depression," and forms with the three lakes an independent hydrographic system absolutely distinct from that of Lake Victoria.

Thus it is that this Ruwenzori range sheds the waters of all its slopes east, west, north and south into one and the same river basin, feeding almost unassisted the three lakes and the Semliki, which together form the south-western sources of the Nile. Furthermore, Ruwenzori being without doubt the most considerable group of snowy mountains on the African continent, and situated in the middle of that continent, and running in the direction of its main axis, does not form a portion of its main watershed. The actual watershed between the Congo and the Nile consists of a line of low hills, lying at a short distance to the west of the Semliki, and masked by the great Congo forest, running northward along Lake Albert, prolonged southward in the volcanic chain which forms the dividing ridge of the rift, between Lakes Kivu and Albert Edward, and finally skirting the eastern shores of Lake Kivu and of Lake Tanganika.

The Duke of the Abruzzi preserves to the chain the name of Ruwenzori, given to it by its first discoverer, Stanley, and adopted since then by the majority of geographers.

Stanley had heard the natives dwelling to the north and west of these mountains call them by the names Ruwenzori, Ukonju, Bugombowa, Avuruka, Avirika, Ruwenzuru-ru, Ruwenjura, etc. He was of opinion that

Chapter VII.

Ruwenzori was the name most commonly in use in Bantu, and that it was to be translated as "King of the Clouds," or "Rain-maker."

Stuhlmann gathered the names Ru-nssoro and Ru-ndjuru from the Wanyoro and Wakonjo natives. In their dialects Niuru and Nssoro signify rain, which confirms Stanley's interpretation of the name but with a considerable change in the orthography. David also wrote Ru-nssoro; he also heard the highest peaks called Kokora.

Scott Elliot gives Runsororo as the native name, and says that he heard from many sources the name Kiriba, which would mean "high peak."

According to A. B. Fisher, the natives of Uganda have no collective name to indicate the entire range, but only separate names for the individual peaks. He gives, however, the names Rwenzozi and Rwenseri, which he interprets as meaning "Mountain of Mountains" or "*The* Mountain" *par excellence,* or as "The Mountain off there," indicating direction. Birika which resembles the Virika of Casati, and Avuruka and the other variations of Stanley, would simply mean "snow."

Sir Harry Johnston heard the snowy portion of the chain called Euchurru by the Nyoro natives; as Ansororo (snow) in Lukonjo; while among the Southern Bakonjos he gathered the name Obweruka; among the Banjoro, Ebirika; among the Baamba, to the north-west of the range, Gusia; among the Baganda, Gambaragara, etc.

In so great an uncertainty as to the nomenclature, Stanley had every right to select a name, and even if his transcription should not be held to correspond with euphonic exactitude to the native word, yet it might be best to keep

Formation and General Features of Ruwenzori.

it as it stands, even were it only out of respect to the great explorer. After all, had Stanley given the range a name which had nothing to do with the native names, had he called it, for example, " Mountains of the Moon," or " Mountains of Ptolemy," or " Victoria Mountains," or any similar name, all geographers would have accepted his choice without discussion and without any attempt to modify it.

These brief remarks upon the name of Ruwenzori will suffice to indicate the impossibility of attempting to gather local native names for each special mountain and peak of the range. So far similar attempts have given as a result a separate nomenclature for each explorer. It is extremely probable that the natives never had individual, specific names for each peak, all the more so if we reflect that in our own European Alps, many peaks received their name only after the advent of Alpine climbing.

It was clearly indispensable to give to the Ruwenzori range some sort of nomenclature, which is the only means of translating into current language the topographical survey of a region.

Out of natural courtesy towards those of his predecessors who had already christened some of the mountains, the Duke, after his return from Africa, interviewed Sir Harry Johnston and Dr. Stuhlmann upon this subject. An agreement was easy, because both of these great authorities shared the opinions of the Prince, who proposed to give to these mountains the names of travellers long associated with the history of Central African exploration, and confining to single peaks those names which Stuhlmann had given to whole portions of the range.*

* Sir Harry Johnston had already suggested that the mountains should be called by the names of celebrated explorers in those cases where no precise and specific native names were forthcoming. (*See* " The Uganda Protectorate," 2nd Ed., London, 1904, Vol. I, p. 159.)

Chapter VII.

The map of Ruwenzori is the chief geographical result of the Italian Expedition. This map was drawn up from data consisting of numerous angular measurements carried out by the Duke from the different peaks by means of the prismatic compass, which were completed by the mensuration of a base line of 300 yards taken on ground near to Bujongolo, by Commander Cagni, and by him connected with Kiyanja (the Edward Peak of Mt. Baker), and with a rocky peak (Cagni Peak) situated to the north-east of Bujongolo, from the summit of which Cagni himself took all the angles of the peaks with a field theodolite.

Last of all, the calculation of the longitude and latitude of Bujongolo permits us to put the chain in its place on the map of Africa.*

The observations were taken in unfavourable atmospheric conditions, nor was it possible in all cases to take them with an instrument of precision so heavy, delicate, and cumbersome as the theodolite. In spite of these drawbacks the topographical sketch may be considered as fairly accurate, because it is based upon numerous observations often repeated over and over again at the same points, and which are in great part reciprocal, so as to admit of mutual verification one from another.

The range of Ruwenzori is situated less than half a degree north of the equator, and about 30° long. E. Greenwich.

---

* According to the map annexed to this volume, Bujongolo is situated at 0° 20′ 23″ lat. N., and 30° 1′ 34″ long. E. Greenwich. The numbers are a few seconds above those given on the map which accompanies the lecture of H.R.H. before the Italian Geographical Society, and published in Fasc. 2, Part 2, Vol. VIII of the "Bollettino," because it was only later that he obtained from the Astronomical Observatory of Greenwich the necessary data for the correction of the lunar tables contained in the ephemerids in order to assign its exact value to the straight ascent of the moon.

Formation and General Features of Ruwenzori.

The general direction is north and south, and the shape is very nearly that of a written G. The principal groups would compose the main curve of the G, while one group only, that farthest south, would represent the tail of the letter.

The range consists of six mountains, *i.e.*, groups of peaks with glaciers, divided from one another by cols without snow, and therefore quite clearly distinct from one another. The area actually covered by glaciers is a little more than seven miles long in a straight line from south to north, and about four miles wide from east to west. The length of the watershed ridge, including all the groups, that is to say, the entire snowy range, is about 11 miles long.

The chain begins in the north with two groups, two parallel snow ridges running nearly due north and south. The easternmost of these was named by the Duke Mt. Gessi, in memory of the Italian explorer who was the first to circumnavigate Lake Albert. The western group was named Mt. Emin, after Emin Pasha, who traversed the Semliki Valley for the first time with Stanley.

Mt. Emin joins Mt. Speke, which bears the name of the first discoverer of the sources of the Nile in Lake Victoria. After Mt. Speke the chain turns westward, rises to the highest group, rightly called Mt. Stanley, and sweeps around in an eastward curve to the group which bears the name of Baker, the discoverer of Lake Albert, who had the first glimpse of the mountain ranges of Ruwenzori.

Last of all, the group to the south of Mt. Baker, which runs from north-east to south-west, had been called by H.R.H. Mt. Thomson, in memory of J. Thomson, whose work in Nigeria is well known. But after his return to Europe the Duke was forced to yield to the proposal of the English Geographical

Chapter VII.

Society, which desired that his own name should be in some way connected with his discoveries, and that Mt. Thomson should be called Mt. Luigi di Savoia. The name of Thomson has been preserved to indicate one of the glaciers of the same group.

MT. STANLEY FROM THE EDWARD PEAK OF MT. BAKER.

Mt. Stanley is the group which includes the highest peaks of all, namely, Margherita (16,815 feet), Alexandra (16,749 feet), Elena (16,388 feet), and Savoia (16,339 feet). There is a fifth peak, Moebius, between Elena and Alexandra, and somewhat lower than Savoia, of which the height has not been measured. The so-called " western-most summit " of Mt. Stanley, mentioned by Freshfield,* and visible from Butiti,

* *See* " Geog. Jour.," 29th March, 1907, p. 327.

MOUNT STANLEY

1   Savoia   Peak
2   Alexandra   ,,
3   Margherita   ,,

appearing conspicuously to the right of Margherita Peak on the panorama taken by H.R.H. from Mt. Gessi, is not so much a real peak as a projecting shoulder upon the north-west ridge of Margherita Peak as may be clearly seen on the plate facing p. 178. Next in height comes Mt. Speke with its two peaks, Vittorio Emanuele (16,080 feet) and Johnston (15,906 feet). Mt. Baker follows with Edward Peak (15,988 feet) and Semper Peak (15,843 feet), which were the first climbed by the

NORTH-WEST SHOULDER OF MARGHERITA PEAK.

Duke. To the east of these are Wollaston Peak (15,286 feet), named after Dr. Wollaston, who was the first to ascend it, and Moore Peak (15,269 feet). The rocky spur on the ridge at the top of the Mobuku Glacier has preserved the name of Grauer who was the first to discover it.

Of the two northernmost groups Mt. Emin includes

Chapter VII.

Umberto Peak (15,797 feet)* and Kraepelin Peak (15,752). Mt. Gessi includes Iolanda Peak (15,647 feet) and Bottego Peak (15,483 feet). Last of all, Mt. Luigi di Savoia includes the Peaks Weismann (15,299 feet), Sella (15,286 feet) and Stairs (15,059 feet). The rocky point, opposite Bujongolo (14,826 feet), is named after Cagni, who climbed it to complete from it the triangulation.

The principal glaciers have taken their names from the peaks from which they flow down.

The five passes which separate the six groups from one another have been named, proceeding from north to south, Roccati, Cavalli, Stuhlmann, Scott Elliot and Freshfield. They are all above 14,000 feet in height, except Stuhlmann's Col between the two principal groups, Speke and Stanley, which is only 13,757 feet high.

The Duke of the Abruzzi has left the native names used by the Bakonjo to the valleys, lakes, rivers and torrents. When, however, the names were numerous, he took no account of them ; and he also left without name those valleys, lakes and torrents on the western slopes which were unknown to the Bakonjo. It remains for some future explorer to gather the native names from the western tribes.

---

* Umberto Peak is 15,797 feet high and not 15,907 as was printed by error in the map reproduced from that of the Italian Expedition by the Royal Geographical Society, and published with H.R.H.'s London lecture in the "Geographical Journal" for February, 1907. The same map also assigns to Moebius Peak of Stanley an altitude of 16,214 feet. This altitude is merely approximate, because no barometrical observation was taken on Moebius Peak. Also the altitude of Weismann Peak is 15,299 feet and not 15,273 feet. I take this opportunity to note that the highest peak of Mt. Baker and the second peak of Mt. Stanley are to be called simply Edward Peak and Alexandra Peak, not *King* Edward and *Queen* Alexandra ; by analogy with the names Margherita, Vittorio Emanuele, and Elena Peaks, and also with the nomenclature adopted in the neighbouring regions, *e.g.*, Lake Victoria, Lake Albert, etc.

RUWENZORI RANGE

of the Valley System of the

Semliki R.

Butagu R.

Russirubi R.

Bujuku R.

Mt Emin

Margherita Pk.
16815

Mt Stanley

C

D

E

Mt Baker

Mt Speke

Mt Gessi

B

Bujongolo

G

F

Mt Luigi di Savoia

H

Kichuchu

Mt Portal

Nakitawa

Mahoma R.

Bihunga

Ibanda

Mobuku R.

Bujuku R.

Buamba R.

Scale 1:278,200 (4.38 miles to 1 inch)

1/2 0   1   2   3   4   5   6   7 Miles

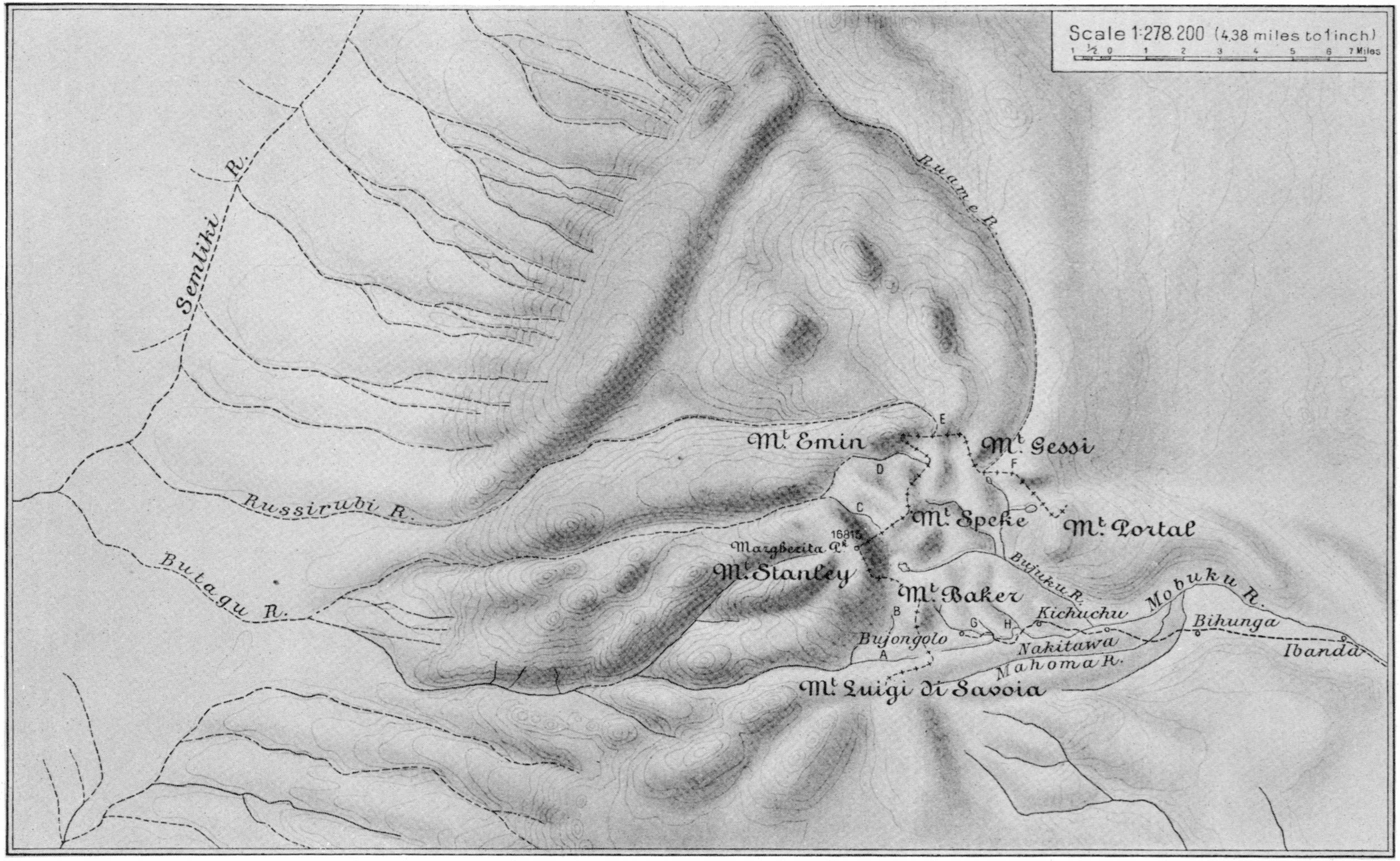

Formation and General Features of Ruwenzori.

The watershed line starting from the peak to the extreme south, the Weismann, runs eastward along Mt. Luigi di Savoia, then northwards over Freshfield's Col and along Edward Peak and Semper Peak of Mt. Baker. From this point, making a wide half-circle, it runs along the high ridge of the Bujuku Valley over Scott Elliot's Col, over the peaks of Mt. Stanley and Stuhlmann's Col as far as Vittorio Emanuele Peak, then descends along the north-east ridge of Mt. Speke to Cavalli's Col, traverses the summit of the two parallel groups, Emin and Gessi, crossing Roccati's Col between them. From Iolanda Peak of Mt. Gessi it follows a south-easterly ridge to the group of the Portal Peaks, whence it turns north-east again.

The most important river basin to the east of the chain is that of the Bujuku Valley, which is surrounded by five mountain groups and is fed by the greater glaciers of Mt. Stanley, Mt. Speke, and Mt. Gessi. The Upper Mobuku Valley, on the other hand, receives only the waters of the Baker glaciers and of a few little glaciers on the eastern side of Mt. Luigi di Savoia through the Mahoma torrent. Hence the Mobuku River is much smaller than the Bujuku River, and is in reality a mere affluent of it. It would, therefore, be more geographically correct to name the entire valley Bujuku, even if only because the two greater mountains stand at its head as well as Stuhlmann's Col, which is the deepest depression of the whole range and lower than Freshfield's Col. The name Mobuku, however, as applied to both river and valley, has been so widely spread by preceding explorers that H.R.H. did not think fit to change it, in order to avoid confusion in the nomenclature.

The atmosphere around Ruwenzori is so misty and so lacking in transparency, even in fine weather, that the Duke

Chapter VII.

never succeeded in getting from the peaks a clear view of the valleys to the west of the chain so as to obtain an accurate idea of their direction and distribution. As far as he was able to observe, he formed the opinion that the four valleys running down from the Cols Freshfield, Scott Elliot, Stuhlmann, and Cavalli (marked A, B, C, D on the map) joined together to form the Butagu Valley, which would consequently collect the waters of the western glaciers of Mts. Luigi di Savoia, Baker, and Stanley, and a great part of those of the Speke Glacier and of the glaciers of Mt. Emin. Thus this would be the most important of the western valleys. It is probable that Mt. Emin and Mt. Gessi contribute to feed the Russirubi and the Ruame Rivers (E and F of the map), which, like the Butagu, are affluents of the Semliki, and that the southern valley, Nyamwamba, runs up as far as the glaciers of Mt. Luigi di Savoia. The torrents Yeria and Wimi would not be fed by glaciers at all.

With the help of our precise knowledge of the range we may now attempt to collate with one another, and with the data furnished by H.R.H., the discoveries and descriptions of the preceding explorers.

Of all these predecessors, Stanley was the one who had the most frequent opportunities of seeing either the single peaks or the range from the north, the west, and the south. He left, however, vague records only, and clearly the reality of the picture has been too greatly altered by the illustrator of his book to make it possible to determine the individual mountains in his illustrations. At the very most it is possible to recognize Margherita Peak and Mt. Speke confused in a single group in the view taken from Kavalli to the north of the mountains, and reproduced on p. 230 of " In Darkest Africa," Vol. II.

Formation and General Features of Ruwenzori.

The mountain of which Stairs caught a glimpse on his way up a valley to the north-west of the chain, perhaps the Russirubi Valley, was probably Mt. Emin. This mountain is illustrated on p. 256 of the above-mentioned volume, and this view, taken from the west, corresponds fairly in appearance to Mt. Emin from the east in the photographic panorama taken by H.R.H. from the Iolanda Peak of Mt. Gessi. (*See* illustration, p. 241.)

As to the "Saddle Peak" of Stanley, it certainly corresponds to the two peaks Alexandra and Margherita, which stand in a line running north-east to south-west. To any one observing them as Stanley did from the north-west or south-east, they would appear as twin peaks, whereas one would be hidden by the other if the observer stood in a line with them.

More exact accounts of the chain have been given to us by Stuhlmann. The more or less schematic illustration of the chain, as seen from the southern part of the Semliki Valley to the south-west of the great peaks, and reproduced on p. 281 of Stuhlmann's book,* can easily be identified on the map of H.R.H. There is no doubt that the central and greatest mountain group given by Stuhlmann under the name "Semper" is to be identified with the Mt. Stanley; hence, the two mountains to its right, designated by Stuhlmann with the names Weismann and Moebius, cannot be anything but Mt. Baker and Mt. Luigi di Savoia. As to the mountain called by Stuhlmann Kraepelin, whose summits are barely visible at a considerable distance from Semper, this must be Mt. Emin. From Stuhlmann's point of observation, Mt. Speke must have been hidden by Mt. Stanley, or only partly visible, and easily confounded with it. This fact, namely, the omission of Mt. Speke

* Dr. F. Stuhlmann, "Mit Emin Pasha ins Herz von Africa," Berlin, 1894,

Chapter VII.

from Stuhlmann's diagram, became later on, as we shall see, one of the chief causes of confusion in the attempt to collate the view of the chain from the east with Stuhlmann's description of the chain as seen from the west.

MT. STANLEY FROM THE WEST, FROM A PHOTOGRAPH TAKEN IN THE BUTAGU VALLEY BY DR. F. STUHLMANN.

By the kind permission of Dr. Stuhlmann and his publishers we are able to reproduce the very interesting and fine engraving which faces p. 188 of his book, and was made from a photograph which he took from the highest point reached by him in the Butagu Valley, a hill 13,326 feet above the sea, and separated from the glaciers by a depression containing a little lake. In the sequel, we shall hear of the excursion made by Sella on the glaciers of the western slope, descending from the col in the centre of Mt. Stanley between Moebius Peak and Alexandra Peak. In the course of this excursion he was able

Formation and General Features of Ruwenzori.

to take several photographs from the western slopes, which, when compared with this plate of Stuhlmann's book, leave no doubt as to its representing the western slopes of Mt. Stanley. It shows, proceeding from left to right, the long snowy ridge which forms the characteristic north-western shoulder of Margherita Peak, which is hidden behind the vast cone of Alexandra Peak. Vertically below the ridge to the right of Alexandra

WESTERN SIDE OF ALEXANDRA PEAK.

Peak, at the foot of the glacier, may be seen a sharp, rocky point, which was climbed by Sella in the course of his photographic expedition. Alexandra Peak is succeeded in this

207

plate by Moebius Peak, then Elena Peak and Savoia Peak, with the little tooth of rock between them, which is also clearly visible from the east.

With relation to the point reached by Stuhlmann, and from which this photograph was taken, Brix Förster, in an article* in which he attempts to collate the preceding explorations of Ruwenzori with that of the Duke of the Abruzzi, is of opinion that this point was near the little lakes to the west of Mt. Baker, in sight of the valley traversed by the Italian expedition to climb to Scott Elliot's Col. A mere glance at the map makes it

MOEBIUS PEAK FROM THE WEST.

quite plain that it was impossible that from any point situated so far to the south of Mt. Stanley the peaks should appear as they do in Stuhlmann's photograph. On the other hand, it is probable that the little lake Kigessi-Kissongo, which Stuhlmann saw between himself and the mountains was one of those drawn in the map to the west of Mt. Stanley under

* *See* in "Globus," Vol. XCI, 1907, p. 345.

Formation and General Features of Ruwenzori.

Point Moebius. In fact, looking from this point towards the chain, Alexandra Peak must have nearly entirely covered Margherita Peak, while the Moebius, Elena and Savoia Peaks must have been visible nearly straight in front, as they are shown in the plate. Brix Förster's article contains other inaccuracies. From the upper valley of Butagu, Stuhlmann saw no other mountains beyond the two photographed by him, nor is there any mention in his book of a third mountain beyond, of which he had caught any glimpse in the opening between

THE FOOT OF THE GLACIERS FLOWING WEST OF ALEXANDRA AND
MOEBIUS PEAKS.

the two. The interpretation of Moore's ascent is entirely erroneous, nor did Moore see from the ridge the other mountain groups, as the author states. The valley ascended by David could not be the Russurubi, for the Russurubi does not lead to any col near 16,000 feet. Finally, Brix Förster is mistaken in writing that Dr. Wollaston ascended the Semper Peak of Mt. Baker. He also states in his article that the highest peaks of Ruwenzori are rocky.

Chapter VII.

We are able also to reproduce Stuhlmann's photograph from the same point of another mountain which he believed to be adjacent to, and to the south of Mt. Stanley, the mountain which he had called Weismann and which, as we have seen, corresponds to Mt. Baker. In reality, this

MOUNT LUIGI DI SAVOIA SEEN FROM THE UPPER BUTAGU VALLEY.
(After the photograph by Dr. F. Stuhlmann.)

mountain was completely hidden from his sight by the southern spurs of Mt. Stanley, and the mountain in his plate must be the Mt. Luigi di Savoia, the very same which he had called Moebius.

The first description of the appearance of the mountains from the east is given by Moore, who seems to have been the only one before H.R.H. to have seen the mountains from the Wimi Valley. Moore had, however, a far better opportunity for observation because, being obliged to descend towards the

Formation and General Features of Ruwenzori.

plain, and to go further from the mountains to seek for a ford across the River Wimi, which was swollen, he was able to see the entire chain, not only the single mountains. This view is reproduced in a plate in colours placed opposite the frontispiece of his book.* This plate contains in the middle, and plainly recognizable, Mt. Stanley and Mt. Speke. To their left stretches a snowy ridge of uncertain outline and long enough to comprise the peaks of Mt. Baker and those of Mt. Luigi di Savoia. To the right, separated from Mt. Speke by a wide interval, is another snow peak, Mt. Gessi.

On reaching the Mobuku Valley and ascending it as far as

SAVOIA PEAK TAKE NFROM ALEXANDRA PEAK; MT. LUIGI DI SAVOIA
IN THE BACKGROUND.

Bujongolo, where he established his first camp, Moore fell into an error, which was subsequently shared by all his successors from Sir Harry Johnston to Dr. Wollaston, and which became the chief cause of the uncertainty which reigned up to the Duke's exploration as to the position of the peaks. This error

* J. E. S. Moore, "To the Mountains of the Moon," London, 1901.

211

Chapter VII.

consisted in his belief that at this point in the Mobuku Valley he was in the midst of the highest mountains of the chain which he had already seen from the plain at the foot of the Wimi Valley, and he still further increased the confusion by attempting to identify them with those described and identified by Stuhlmann from the western slope.

It is not easy to make out Moore's ascent. Upon an attentive perusal of his narrative, collated with H.R.H.'s map, the reader is led to suppose that on reaching the head of the Mobuku Valley he started to ascend to the left (that is to say, on the right slope of the valley) until he reached the glacier which he calls the central glacier, in other words the Baker's Glacier of H.R.H.'s map,* by which glacier he would reach the ridge at a point between Semper Peak and Grauer's Rock. As a matter of fact, however, in order to reach the Baker Glacier from the valley it would be necessary to climb rocks and gullies presenting such exceptional difficulties as to be surmountable only by a party of trained mountaineers—certainly not by a single white man accompanied by native porters. It is more probable that Moore began to climb the right slope of the valley at an earlier point. In this way he would have reached the Edward Glacier and ascended it to the southern ridge of the Edward Peak.

Sir Harry Johnston attempted to reconstruct the chain as seen from a hypothetical point to its east, basing his conception upon the observations taken by preceding explorers. The representation thus obtained by him is much further from the truth than that of Stuhlmann and of Moore. From the

---

* The glacier is clearly shown in one of Moore's illustrations (p. 246), and also in a plate of Sir Harry Johnston's, " The Uganda Protectorate," 2nd Ed., London, 1904, Vol. I, p. 178.

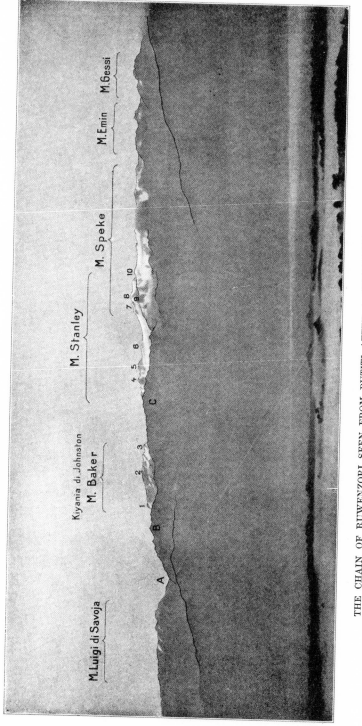

THE CHAIN OF RUWENZORI SEEN FROM BUTITI AFTER A TELE-PHOTOGRAPH BY V. SELLA.

A. Freshfield Col.
B. South Portal Peak.
C. North Portal Peak.

1. Knob on south ridge of Mt. Baker ascended by
   Wollaston.
2. Edward Peak.
3. Semper Peak.
4. Savoia Peak.
5. Elena Peak.

6. Moebius Peak.
7. Alexandra Peak.
8. Margherita Peak.
9. Johnston Peak.
10. Vittorio Emanuele Peak.

Formation and General Features of Ruwenzori.

Lower Mobuku Valley he saw a great snowy mountain which he named Duwoni. He gives a fine illustration of this mountain on p. 158 of his book. There is no doubt that this is to be identified with Mt. Speke. On reaching Bujongolo he believed himself to have reached the foot of this same Duwoni, whereas in reality he had been going further from it from Nakitawa onward. He furthermore believed that Mt. Kiyanja (Baker) was Mt. Semper of Stuhlmann, and that Duwoni (Speke) was Peak Weismann.

Mr. Freshfield, as well as the Duke, was able to have a complete view of the range from Butiti, on the way between Entebbe and Fort Portal. He enumerates * the mountains of Ruwenzori as follows, from left to right :—1st, a massive rocky group with patches of snow, which he calls South Peak, and which is Mt. Luigi di Savoia ; 2nd, a wide col, which is the col above the Mobuku Valley which now bears his name ; 3rd, a bold peak of rock and glacier, the Kiyanja of Johnston ; 4th, the undulating ridge covered with glacier which from this last group leads to the highest snow peak, and which Mr. Freshfield takes to be the Duwoni of Johnston, but which is in reality Margherita Peak. Duwoni or Speke, as a matter of fact, does not appear as an isolated mountain when seen from Butiti, but seems to form a single group with Mt. Stanley. It may be of use to the reader at this point to compare Mr. Freshfield's description with the outline of the range as seen from Butiti, taken from Sella's tele-photograph, and with the addition of the new names. On reaching the head of the Mobuku Valley, Mr. Freshfield would appear to have fallen into the same error as his predecessors, for he describes it as

---

* D. W. Freshfield, ' A note on the Ruwenzori Group,' " The Geographical Journal," May, 1906, Vol. XXVII, p. 481.

Chapter VII.

' enclosed in a cirque of cliffs capped by glaciers, which flow from a névé rising in comparatively gentle slopes to an icy ridge connecting two rock peaks, Kiyanja and Johnston's Duwoni."*

Mr. Freshfield's poor opinion of the glaciers and of the general importance of the chain is a natural result of his belief that " the only glacier basin of any size east of the chain is that of Mobuku."†

It now remains for us to consider the ascents performed by the members of the British Museum Expedition, and more especially by the mountaineer, A. F. Wollaston. Lake Bujuku seems to have been seen for the first time by Woosnam in the excursion which he made alone to the ridge overhanging the Mobuku Glacier. Mr. Woosnam, as well as Mr. Wollaston, believed, however, with Dr. Grauer, that this was the water-shed. Hence Mr. Wollaston naturally concluded that the mountains which he had caught a glimpse of beyond it, Mt. Stanley and Mt. Speke, were on the western slopes of the chain. It was only later, after meeting H.R.H. at Fort Portal, that Mr. Wollaston, while crossing the foot of the Mobuku Valley, and seeing the outline of the peaks to the west, finally understood that their eastern slopes do really form part of the Uganda side of the range.

I must here mention the interesting article of Lieutenant T. T. Behrens,‡ who has attempted to reconstruct the chain of Ruwenzori with the whole illustrative and descriptive material from Stanley to Wollaston which existed in July, 1906, including the observations taken by the author during nine

---

* D. W. Freshfield, in " Alpine Journal," August, 1906, p. 183.
† D. W. Freshfield, in " Alpine Journal," August, 1906, p. 201.
‡ Lieutenant T. T. Behrens, ' The Snow Peaks of Ruwenzori,' " The Geographical Journal," July, 1906, Vol. XXVIII, p. 43.

Formation and General Features of Ruwenzori.

months of residence in the regions close to the chain during the work of the Anglo-German Boundary Commission. Behren's article contains in clear and succinct shape the net result of all that was known about Ruwenzori previous to the Italian Expedition.

The following table, in which the names given to the mountains by different travellers are placed in order with reference to those marked upon the Italian map, illustrates the confusion which had arisen from mistaken identification of peaks from various points of observation. This table will make it easier to compare the accounts of all the previous journeys. It will also show that the only way to put order and clearness into the nomenclature was to give up the old names and start afresh with a different plan.

As regards the altitude of the highest point as determined by the Duke at 16,815 feet, it would be idle now to gather together the opinions of preceding explorers which were not based upon any instrumental observation ; all the more so that only one of them, Stuhlmann, really had a near sight of the highest peaks, or at least of Alexandra Peak. The others all judged of the height of the peaks around the head of the Mobuku Valley, and never even saw the highest ones, or only from the far-off plains of the surrounding country.

The Duke took for the first base of his calculations the meteorological station of Entebbe, whose height was already known (3,861 feet). Fort Portal was then connected with Entebbe by a series of observations carried out for two whole months at the two stations, and giving as a result for Fort Portal an altitude of 1,165 feet above Entebbe, or 5,026 feet above the sea-level. Finally, Bujongolo was connected with Fort Portal by barometrical observations taken during a period of about a

Chapter VII.

THE NOMENCLATURE OF RUWENZORI AND THE

H.R.H. the Duke of the Abruzzi.	F. Stuhlmann.		J. E. S. Moore.	
	From Lungwe (Semliki).	From the Butagu Valley.	From the Eastern Plain.	From Bujongolo.
Mt. Luigi di Savoia.	Moebius.	Weismann or Ngemwimbi.		
Mt. Baker. Edward and Semper Peaks.	Weismann or Ngemwimbi.		Moebius.	Ingomwimbi.
Mt. Baker. Wollaston and Moore Peaks.	Weismann or Ngemwimbi.		Moebius.	Kanyangogwe.
Mt. Stanley. Savoia and Elena Peaks.				
Mt. Stanley. Alexandra and Margherita Peaks.	Semper or Kanjangungwe.	Semper or Kanjangungwe.	Ingomwimbi.	
Mt. Speke.			Kangangogwe.	
Mt. Emin.	Kraepelin.			
Mt. Gessi.			Saddle Mountain (of Stanley).	

Formation and General Features of Ruwenzori.

ERRORS IN THE IDENTIFICATION OF THE PEAKS.

Sir Harry Johnston.		D. W. Freshfield.		A. F. Wollaston.
Figurative Scheme.	From the Mobuku Valley.	From Butiti.	From Bujongolo.	From Bujongolo.
		South Peak.		
	Kiyanja.		Kiyanja.	Kiyanja.
	Duwoni (from Bujongolo).	Kiyanja.	Duwoni.	Duwoni.
Kiyanja (Semper of Stuhlmann).				
		Duwoni (Semper of Stuhlmann).		
Duwoni (Weismann of Stuhlmann).	Duwoni (from the lower valley).			

Saddle Mt. (?)

Chapter VII.

month, from the 16th of June to the 12th of July. In this way the altitude of Bujongolo was established at 7,435 feet above Fort Portal, consequently 12,461 feet above the sea-level. The observations taken in the valleys and on the peaks with the mercurial barometer, or in places of secondary importance with the aneroid, were then referred to this base. A third base of less value, because the observations there were not taken during so long a period as at Bujongolo, is Ibanda in the Lower Mobuku Valley. Only one of the measurements of altitude refers to this as its base point, namely, that of the Iolanda Peak of Mt. Gessi, which was also taken with the boiling-point thermometer because the barometer was broken.

To estimate the approximation to the actual truth, which may be attained by measuring heights merely by direct observation of atmospheric pressure, by means of the boiling point of water (hypsometer), or by the aneroid barometer, it will suffice to compare some results obtained by this simple method with the corresponding measurements as taken by the Duke and calculated with all the corrections furnished by a base station.

Locality.	Grauer, Tegart, and Maddox.		Wollaston.	H.R.H. Observations referred to a base station.
	Hypsometer.	Aneroid.	Hypsometer.	Mercurial Bar.
Fort Portal ...	5,200	—	—	5,027
Bihunga	6,978	6,700	—	6,300
Kichuchu ...	9,869	9,600	—	9,833
Bujongolo ...	12,481	12,300–12,500	12,660	12,461
Camp Grauer ...	13,303	13,100	—	13,229
Grauer Rock ...	14,956	19,030	—	14,813
Wollaston Peak ...	—	—	15,893	15,286

Formation and General Features of Ruwenzori.

By far the most interesting altimetric observations are those which were taken by Lieutenant T. T. Behrens in 1903 during the Anglo-German frontier delimitation. He then fixed trigonometrically the height of the two peaks which appeared to be the highest, as well as of four other points. It was, however, only after the publication of the Italian map that he was able to identify with certainty the position of these peaks, and hence to obtain all the data for an exact calculation of the altitudes on the basis of the angles taken. It is worth while to compare his results with those of H.R.H.

Name of Peak.	Height in Feet above the Level of the Sea.		
	Lieut. T. T. Behrens. Trigonometric Value.	H.R.H. Barometric Value.	Difference, Trigonometric *minus* Barometric.
Margherita	Feet. 16,619	Feet. 16,815	− 196
Alexandra	16,543	16,749	− 206
Edward	15,748	15,988	− 240
Vittorio Emanuele ...	15,846	16,080	− 234
Umberto	15,554	15,798	− 244
Iolanda	15,258	15,647	− 389

This is not the place to discuss the relative value of the two series of numbers. In Appendix B, Prof. Omodei gives *in extenso* all the data of H.R.H.'s observations, and a critical *exposé* of the methods adopted, and of the precautions taken in calculating the altitudes.

Dr. A. Roccati has written a detailed account of the geology and mineralogy of Ruwenzori, which is published in the volume

Chapter VII.

containing the work done by the Italian expedition in different branches of natural science.*

Like Stuhlmann and Scott Elliot, he absolutely excludes all possibility of a volcanic origin for the chain. There is

SENECIO AND HELICHRYSUM IN THE UPPER BUTAGU VALLEY.

only one circumscribed point of the whole region explored, namely, the rock wall which forms the shelter of Kichuchu, where a formation of a volcanic nature exists, consisting of veins of basalt enclosed in a wall of gneiss.

The inclination of strata is often very marked, reaching at times an angle of over 60°, and is, as a rule, turned eastward

---

* Appendix D contains a short résumé by Dr. Roccati of the geology of the range.

Formation and General Features of Ruwenzori.

and south-eastward on the eastern part of the chain, southward on the southern part, and south-westward on the western part so as to form a tectonic semi-ellipse.

Resuming in its main outlines the origin of the group of mountains and of the high peaks of its central portion, we may ascribe it to three causes, geotectonic, stratigraphic and lithologic, namely:—

1. The upheaval *en bloc* of a whole portion of the archaean rocks of Central Africa with a main slope for the rise from west to east. This rise is mainly in relation to the gigantic western fracture, with its relative vertical displacements, which originated the Semliki Valley, and also with other fractures which have taken place to the east of the group, and which are marked by a series of recent volcanoes like those of the Province of the Toro District.

2. To a marked elevation—ellipsoid or anticlinal, with general direction from north to south, and strata more or less considerably uplifted in the Ruwenzori group.

3. To the existence in the central region of a group of rocks which have resisted the physico-chemical action of external agents (amphibolite, diorite, diabasis, amphibolic gneiss), whereas the gneiss-rocks and mica-schists of the lower zone oppose far less resistance to such agents.

To these main causes we may add the probable existence of internal fractures in the group, with a main direction from north to south, which would have contributed to the isolation of the several groups. An important geological feature of Ruwenzori is the vast development of the glaciers during the glacial period. The proofs of this are manifest, especially at Nakitawa. At one time the valleys of the Mahoma, of the Mobuku and of the Bujuku were filled with large glaciers

223

which met below Nakitawa and flowed down to a point beyond Bihunga. In this way also, on the western slope, the Savoia, Elena and Semper Glaciers must have entered into the hollow formed between the groups Stanley, Baker and Luigi di Savoia,

LOBELIA DECKENI, SENECIO AND TREE-HEATHS.

and probably joined the Edward Glacier. It is not possible to state exactly down to what point the glaciers had reached in that direction, as the valleys to the west of the chain were not explored.

224

Formation and General Features of Ruwenzori.

To-day the glaciers are of small extent and diminishing. This is proved at some points by the presence of moraines recently abandoned only a few hundred yards from the actual glacier snout, and from the freshness of the marks of polishing by ice on the rocks close to nearly all the glaciers. There are no glaciers of the first degree in the principal valleys, but only secondary glaciers in the upper part of the mountains and in the main gorges, not, however, in the nature of mere hanging glaciers, but true glaciers. Unlike our own Alps, there are no real basins, but merely a sort of glacier caps from which ice digitations flow down at divers points. In other words, we have on the higher groups of Ruwenzori glacier formations which remind us of the Scandinavian type and which have been called tropical glaciers.

The Moore and Semper Glaciers flow further down than any—the former as low as 13,690 feet, and the latter as far as 14,000 feet. The largest glaciers are on the Stanley, Speke and Baker groups, and on the eastern sides of the Gessi group. The smaller ones are upon the Emin and the Luigi di Savoia groups, unless these latter have important glaciers to the north of the one and the south of the other where they were not explored by the expedition.

A characteristic feature of the high ridges, and more especially of the snowy ridges around Alexandra and Margherita Peaks, are the enormous cornices, which from a distance appear to be inaccessible, and have a totally different appearance to those of the Alps and of the Caucasus. Rapid and frequent changes in temperature, falling from several degrees above to several degrees below the freezing point, create an incessant alternation of frost and thaw, and give rise to the formation of an immense number of stalactites under these

225

Chapter VII.

cornices, which are so intricate and so situated in relation to one another as to form a real scaffolding to support the ice-dome, which is usually of a spongy consistency and quite light.

LOBELIA IN FLOWER.

Thus on Ruwenzori the cornices are far more solid and safe than in the Alps, and, in spite of their number and extent, there was no visible sign of a recent collapse at any point.

Formation and General Features of Ruwenzori.

The snow-limit may be calculated as between 14,700 and 14,800 feet; in other words, at about the same level which is reached by the lower extremity of most glaciers. Towards 14,000 feet the rain always turns into snow.

There is not, perhaps, an absolute predominance of any one wind throughout the chain. It is, at least, quite certain that fog, snow, and hail are extremely frequent and common to all the winds, so that all forecasts are vain. Fine and bad weather may alternate several times in a few hours, and in so capricious a manner as not to appear subject to any law. Only in the early morning hours there seems to be a somewhat greater probability of a clear sky.

The bad weather is frequently violent, and accompanied by strong wind, lightning and thunder, even in the highest regions. Near Alexandra Peak and on Edward and Sella Peaks the rocks bear witness to the violence of these storms by the innumerable fulgurites with which they are riddled. No conclusions as to the best season for visiting Ruwenzori can be drawn from the experiences of the Italian expedition. During June bad weather certainly predominated. The longest dry spell was in the second week of July. After this period the expedition began to withdraw from the higher valleys and peaks, to which the rains and fogs seem to be strictly limited. Indeed, on their return to Fort Portal they learned from the resident missionaries that during those two months no rain had fallen there.

The valleys of Ruwenzori are often divided into natural terraces produced by the formation of layers of strata above the ridges of hard rock, which at an earlier period dammed up these valleys in places, thus creating lake basins which have subsequently silted up with alluvial deposit, of which the present marshy levels are the result. Lake Bujuku is a survival of one of these ancient basins.

In the Mobuku and Bujuku Valleys towards 10,000 feet the damp and mild climate specially favours the development of

SENECIO AND LOBELIA STUHLMANNI.

lycopodium, mosses, and lichens which clothe the sides and bottoms of the valleys, and cover the trunks of the living trees or of those that have fallen from old age. At this height the valleys are clad with a dense forest of heaths and

228

IN THE SENECIO FOREST

Formation and General Features of Ruwenzori.

of bamboos, with brambles, orchids and ferns, in whose shade grow violets, ranunculi, geranium, epilobium, thistles, and umbelliferous plants.

Towards 11,500 feet a certain number of the aromatic plants, which had formed a large part of the underwood, cease, and among the trees only the heaths, lobelias, and senecios remain, while the ferns become prominent, and the lycopodium, mosses, and lichens develop to an inordinate degree. This development reaches its maximum at about 12,500 feet, a little before the point where the heaths stop altogether, beyond which remain only senecios, lobelia, reeds, mosses, and lichens.

Here the helichrysum, or everlasting flower, which had already been noticed at about 11,500 feet, forms dense thickets, which reach up to the glaciers together with the senecios, and is the last form of shrubby vegetation. Among the numerous specimens of helichrysum and senecio brought home by the expedition, there were several new species. Mention should be made here of a fine and rare tree found at Bujongolo, belonging to the family Ericaceæ, and the genus Philippia.

On the peaks are seen mosses, lichens, a few rare graminaceæ and a few dwarf phanerogams which remind one of the characteristic vegetation of our own Alps. At and above 16,000 feet the rocks are bare.

The expedition had not proposed to itself the task of making special and minute researches regarding the fauna of the region. So far, however, as was possible, in spite of the rapidity of the marches and the unfavourable circumstances, as many animal specimens as possible were collected, and to these were added large collections made for the expedition by the Catholic missionaries.

As they ascended the Mobuku Valley the fauna became less

Chapter VII.

and less abundant, while above Bujongolo nothing was found except leopards, rats, bats, a few crows, hawks, birds of the sparrow family, insects and worms. Upon the peaks were found worms, neuroptera and diptera.

The botanical and zoological collections offered abundant material, comprising many species which were either new or interesting from other points of view, and which have been amply described and illustrated in the volume of special studies upon Ruwenzori.

We have now glanced cursorily at the principal results of the Italian expedition—results implying sustained effort, owing to the extreme shortness of their sojourn in the mountains. Having thus briefly reviewed the chief features of the region, we may once more take up our narrative.

CHAPTER VIII.

EXPLORATION OF MT. SPEKE AND MT. EMIN.

The Descent into the Bujuku Valley—Stuhlmann's Col—Western Slopes of Mt. Speke—Ascent of Vittorio Emanuele Peak—Storm and Electrical Phenomena upon the Peak—Two Days' Bad Weather—Glacier Torrents of Ruwenzori—The Duke reascends Vittorio Emanuele Peak—Crossing the Western Valleys—Camp at the Foot of Mt. Emin—Ascent of Umberto Peak —Return to Bujongolo—Three Days' March through the Rain—Recapitulation of the work done

WE left off the narrative at the point when the expedition had assembled in Camp IV, above Scott Elliot Col, near the Elena Glacier of Mt. Stanley, on the 21st of June.

On the following morning, June 22nd, the Duke once more prepared to leave his companions and pursue the exploration of the chain, directing his steps towards the northern groups. Some days before, from the summit of Alexandra Peak, he had been able to ascertain that the most convenient and the shortest route to Mt Speke and Mt. Emin lay along their western slopes, which could be easily reached by crossing the Upper Bujuku Valley and Stuhlmann's Col,

231

Chapter VIII.

CAMP IV, CLOSE TO THE ELENA GLACIER.

which lies between Mt. Speke and Mt. Stanley. The Duke
was accompanied by the guides Joseph and Laurent Petigax,
Ollier and five Bakonjo porters.

The northern wall of the Scott Elliot Col, as has already
been stated, forms a precipice overhanging Lake Bujuku.
After a short descent between great blocks of rock covered
with the usual mosses and lichens, the party entered a
narrow gorge and then a nearly perpendicular gully full of
detritus. The loaded natives, little used to the precautions
which are necessary under these circumstances, trod carelessly
upon the stones and set them rolling, to the serious risk of
those who were ahead. It was necessary to descend slowly,
with great caution, and quite close together.

The natives had by this time acquired somewhat greater

Exploration of Mt. Speke and Mt. Emin.

confidence in their leaders, and followed them with a better will. The guides helped them at all the difficult points and the caravan was once more in good spirits. At the foot of this gully they again entered the senecio forest, through which they descended by a gentler slope obliquely towards the bottom of the valley, where they found a treeless, marshy tract (12,904 feet). This they reached after two and a-half hours' march, crossing the torrent above Lake Bujuku. The usual obstinate fog enveloped the whole valley.

From this point they began to ascend by a moderate slope directly towards the southern face of Mt. Speke. About 300 feet higher up, they came across a névé formed by avalanches at the foot of the rock wall, which was covered above by a glacier broken up into séracs, and, as it were, suspended over the valley. Skirting round the névé to the left they continued to ascend, directing their steps to a point where the south-west ridge of Mt. Speke joins Stuhlmann's Col to the north of a rocky spur clearly visible in the middle of the col. The last bit of way at the foot of the perpendicular cliff, overhung by the terminal séracs of the glacier, is exposed to the danger of stones falling from above. The ascent is easy but fatiguing on account of the slippery rocks covered with moss.

By the time they reached the watershed, a fine warm sun had dissipated the fog, and they stopped for a while to enjoy the fine view over the upper amphitheatre of the Bujuku Valley. It is completely surrounded by precipitous cliffs. Only under Margherita and Alexandra Peaks the gentler slope allows the glacier to descend to a lower level, but all the rest of the circle of glaciers stops short at the brow of the cliffs. Now and then the roar of the avalanches of séracs may be

Chapter VIII.

heard as they crash down into the valley. The rocky buttresses of Elena and Savoia Peaks, and the precipitous cliffs of the north face of Mount Baker, overtopped towards the east by Moore and Wollaston Peaks, are especially grand.

After skirting the foot of the south-west ridge of Mt. Speke, they pursued their way nearly on a level under the western cliff, keeping high and not far from the glacier. This glacier has withdrawn recently, leaving a long fringe of rocks and moraine detritus, under which a few senecios and groups of helichrysum have taken root.

A little further on, the tent was pitched on a narrow strip of land between two oval lakes and the margin of the Speke Glacier. This is Camp V, at a height of 14,682 feet above the sea-level, immediately under Vittorio Emanuele Peak. There were only a few senecios at this point, and the natives sought for shelter lower down, where there was abundance of wood. The sky was clear overhead, but round the peaks and in the valleys lingered fogs, which hid the greater part of the landscape. A little further and lower down was a third lake, somewhat larger than the two which were near the camp.

On the next day, first climbing the rocks and then up the glacier, following an easy western ridge, without once using the rope, in a little more than an hour they reached the summit of Vittorio Emanuele Peak, 16,080 feet above the sea-level. It was 6.30 in the morning and they were already surrounded by dense fog. They remained nearly eight hours on the summit in vain expectation of an opening in the fog, which never came. There was a light, variable wind, and every now and then a snowfall, changing occasionally into brief and violent showers of hail. At one time they were enveloped in a cloud so charged with

234

SENECIO FOREST TO THE WEST OF FRESHFIELD COL
SAVOIA PEAK IN THE DISTANCE

electricity that tiny discharges began to crackle upon their ice-axes, their tripod, and their barometer. Even their hair

MT. SPEKE SEEN FROM THE SENECIO FOREST AT THE FOOT OF
SCOTT ELLIOT'S COL.

crackled upon their heads. It was a disagreeable situation, and by no means without danger.

To kill time, they built a big stone man on a point of rock to the north-west, a little below the snow peak. In the

Chapter VIII.

afternoon they went down to the camp, as the weather was getting worse. The day had been completely wasted as far as exploring work was concerned.

The 24th and the 25th of June were spent in a dense fog, with alternations of rain, snow, and hail. The guides set out on a short excursion to endeavour to find the way to Mt. Emin across the valleys which run down to the north-west of Mt. Speke. On the 25th, H.R.H. attempted to proceed, but was very soon forced to return, the fog being so dense that it seemed like night.

During these two days of obligatory rest, the Duke was able to observe in the little torrents which flow down from the Speke Glacier the periodic oscillations of volume, ranging from a minimum in the morning to a maximum in the evening, which are characteristic of torrents produced by the melting of ice. There would be no reason to comment upon this fact were it not that Mr. Freshfield was led, from observation of the small dimensions of the Mobuku torrent where it springs from the glacier, and from the limpidity of its waters, to conclude that it rather originated from a spring under the glacier than from actual melting of the ice.

This view fits in with Mr. Freshfield's general theory that tropical glaciers are consumed chiefly by evaporation, and in a minor degree only by melting. Whatever may be the conditions in the Himalayas, there can be no doubt that in this respect the glaciers of Ruwenzori resemble those of our own Alps, and that they all give rise to torrents flowing from their extreme end with all the characteristic features of glacier torrents. As a matter of fact, the climate of Ruwenzori is very little tropical in its nature, and it would seem that a condition of atmosphere so saturated with moisture as to render the

Exploration of Mt. Speke and Mt. Emin.

mists nearly permanent could hardly cause so rapid an evaporation as alone to account for the very considerable waste of the glaciers. The limpidity of the waters of torrents which spring from certain glaciers of Ruwenzori may, in all probability, be ascribed to the almost complete immobility of the glaciers themselves, owing to which they grind no detritus from the rocks that form their beds. As was mentioned in the preceding chapter, these glaciers are in the form of ice-caps on the summits and ridges rather than of true streams of ice flowing from névés, as is the case in our own Alps.

Fully to estimate, however, the importance of the Ruwenzori chain in feeding the Nile, we must take into account not so much the glaciers as the entire mountain range, whose highest peaks soar up into the colder strata of the air, and gather to themselves and precipitate in rain and snow the mass of vapours drawn up from the vast plains below, while the network of valleys form great basins to collect the water thus gathered. The reader will remember that on the western and southern slopes alone Stanley counted sixty-two torrents flowing from the mountains into the Semliki River and into Lake Albert Edward.

On the evening of the 25th of June the scene changed rapidly. The whole sky cleared up, and a marvellous sunset kindled the whole valley and the far-off forest of the Congo into flaming red.

The following night was bitterly cold. On the morning of the 26th, the Duke and the guides were on their way by four o'clock. The frost was hard and all the water frozen, even the little lake was nearly completely covered with ice. The hard snow gave a good foothold upon the glacier. By a quarter past five they were once more on the summit of Vittorio

Chapter VIII.

Emanuele Peak. A cold wind was blowing from the north-west. In the absolutely clear and transparent air the outlines of the peaks stood out distinctly upon the sky. The weather was capital for topographical work.

Vittorio Emanuele Peak.   Johnston Peak.

MT. SPEKE FROM THE STANLEY PLATEAU.

Vittorio Emanuele Peak is situated nearly in the centre of the vast circle upon which are distributed the mountains and glaciers of Ruwenzori, and is without doubt the best point of observation of the whole range. Northward from the peak runs the long ice ridge which presently dips to the Cavalli Pass, and thence rises again to the Umberto Peak of Mt. Emin. A deep gorge between precipitous cliffs, running from north to south, divides Mt. Emin from Mt. Gessi. The two peaks of Mt. Gessi, Iolanda and Bottego are clearly visible at the

MOUNT SPEKE FROM EDWARD PEAK, MOUNT BAKER

Exploration of Mt. Speke and Mt. Emin.

extreme end of the terminal snow ridge. To the south-west rises the mighty mass of Mt. Stanley with its five peaks, of which the Alexandra Peak is scarcely visible to the left and to the back of Margherita Peak; while on the great ice plain forming the Stanley plateau they were able to discern, like little black specks, the caravan of Vittorio Sella, on his way up to the Alexandra Peak that morning. The ring of glaciers ends to the east with the Moore Peak of Mt. Baker, where they saw the stone man which Vittorio Sella had set up there a few days before.

Through the opening formed by the Scott Elliot Pass, as if through a window, they saw at a distance the western extremity of Mt. Luigi di Savoia. Between this mountain and the pass the eye follows the valley down to the lakes to the west of Mt. Baker. Here in the pale light of the dawn they made out a fire. This was the camp of the native porters who were bringing supplies.

Thus, at that early hour, from the summit of Mt. Speke, the Duke saw the whole carefully organized work of his expedition proceeding before his eyes.

A little after seven o'clock they were back at the tent, and spent the rest of the day in drying in the sun their equipment, which was soaked with the rain of the previous days.

Towards evening a few more Bakonjo arrived with provisions. The night was absolutely clear and starry, and the sun rose in a perfectly clear sky. The Duke started ahead with a guide and proceeded northward across the interval between the lakes and the foot of the cliff, a rocky ledge upon which, about 600 feet further up, the glacier comes to an end in a cascade of séracs. They proceeded by leaping from one to

Chapter VIII.

another of the blocks heaped at the foot of the cliff. The porters found an easier way a little further down through the senecios and helichrysums between the second and third lakelets.

Thus skirting along the glacier they presently reached the top of the lofty buttress, which runs westward from Mt. Speke, dividing the two valleys which are to the west of the Stuhlmann and Cavalli Passes. A spur of rock rises from the very ridge, forming a *belvedere* 14,744 feet high, from which they were able to observe the way which lay before them.

Unfortunately the weather was already changing, and detached drifts of mist, which had been gathering here and there, now began rapidly to collect and melt into one another. They saw quite clearly from this point a conical rocky peak rising from one of the western buttresses of Mt. Emin—a sort of "little Matterhorn," which may possibly have been one of the "twin cones" towards which Stairs was steering on his expedition to the north-west of the chain.

To reach the foot of Mt. Emin it was necessary to cross the head of the great valley which runs down to the west of the Cavalli Pass and cross another and smaller spur which runs into this valley from Mt. Speke. Hence they continued skirting the mountains at the foot of the Grant Glacier, which seems to have shrunk even more than the others. On reaching the top of this spur, they proceeded to descend, skirting the slope towards the Cavalli Pass, taking advantage of a providential ledge which squeezed a narrow way between smooth steep slabs of rock which would otherwise have been impassable. This ledge was covered with a dense thicket of helichrysum, through which the guides cut a path. The valley was crossed

Exploration of Mt. Speke and Mt. Emin.

near the top, just under the col, and Camp VI was established at a distance of about half a mile below the Umberto Glacier upon a little rocky terrace on the top of a precipice which reached to the bottom of the valley. About 500 feet lower down, this valley forms a sort of amphitheatre surrounded by precipitous cliffs, excepting in the centre where a gentler slope leads to the Cavalli Pass. To the west it narrows

Umberto Peak.   Kraepelin Peak.

MT. EMIN SEEN FROM THE IOLANDA GLACIER.
(*Negative taken by H.R.H.*)

into a gorge through which flows the torrent which springs from the southern glaciers of Mt. Emin. The day's march had been long, and all were tired.

On the morning of the 28th of June there was again a prospect of bad weather. They left with a cloudy sky and proceeded to ascend a rocky ridge which runs down from

Chapter VIII.

Umberto Peak between the Emin and Umberto Glaciers. On reaching the right margin of the latter they left their tent there some 600 feet above the sixth camp and proceeded to ascend over the snow which covered the glacier. At the top they turned westward towards a rocky ridge, by means of which they reached Umberto Peak. Here they remained for five hours, but were scarcely able to catch a glimpse of a peak here and there among the mists. They built a big stone man upon the wide rocky summit.

A great ridge of broken and decomposing rock runs northward towards Kraepelin Peak, which is lower and likewise rocky. Mt. Gessi, on the other side of the narrow gorge, between precipitous cliffs, has the appearance of a vast col with two peaks rather slightly accentuated at the northern and southern extremities of the long snowy ridge. It was late when they returned to the tent near the glacier. The Duke would have liked to return upon Umberto Peak on the following day to take angles, but the weather was threatening from the dawn. They had supplies for one day only. At such a distance from Bujongolo, which was their base, and with so many passes to cross on the way, it was not easy to obtain provisions regularly. It was therefore necessary to return.

In half an hour they reached Camp VI, under the Cavalli Pass, ascended to the *belvedere*, in a snowstorm, and in the early hours of the afternoon set up their tent once more at the foot of Vittorio Emanuele Peak. On the following day, a long march, almost entirely in the rain, brought them over the Stuhlmann Pass, the head of the Bujuku Valley, and the Scott Elliot Pass. They once more set up their tents at Camp II, on the shore of the little lake at the foot of the western slope of Mt. Baker. Rations had been left ready at

Exploration of Mt. Speke and Mt. Emin.

prearranged points on the return route, so that the caravan could move quickly, having only to carry the light camp material.

Finally, on the 1st of July, they crossed the Freshfield Pass, where Vittorio Sella had set up his tent, and was waiting with Botta for fair weather so as to be able to do some work with the camera. The Duke proceeded under falling rain and returned to the muddy Mobuku Valley, and to the camp of Bujongolo after seventeen days' absence.

The Prince had spent the whole of this time at heights above 13,000 feet, with light and barely sufficient equipment, sleeping with his two guides in a single Whymper tent, without a camp bed, with clothes nearly always soaked with rain and snow, and with such discomfort and fatigue as are known only to those who have experienced mountain life under similar conditions.

In the course of these seventeen days he had ascended Margherita, Alexandra (twice), Elena, and Savoia Peaks of Mt. Stanley, Vittorio Emanuele Peak of Mt. Speke (twice), and Umberto Peak of Mt. Emin, crossed the Freshfield, Scott Elliot, and Stuhlmann Passes and explored the head of the Bujuku Valley, and the western slopes of Mt. Speke. He had determined the relative positions of the peaks, and the relation to each other of the several groups, a work already in great part sketched out during his first ascents of the peaks of Mt. Baker, but now completed by numerous altimetric and angular mensurations.

His work was carefully planned to proceed in conjunction with that of the other members of the expedition, in order to insure a thorough exploration of the ranges, as we shall see in the following chapter.

CHAPTER IX.

FURTHER ASCENTS ON MTS. STANLEY, LUIGI DI SAVOIA AND BAKER. WORK AT BUJONGOLO.

Three more ascents of the Alexandra Peak—Ascent of Moebius Peak—Crossing of the Central Col of Mt. Stanley—A week of bad weather on the Freshfield Pass—Ascent of the Edward Peak by the South Ridge—Ascent of the Sella Peak—Work at Bujongolo—Preparation of a Base Line—H.R.H. returns to the Edward Peak—Ascent of the Cagni Peak—Panorama taken from the Edward Peak—Ascent of Peaks Wollaston and Moore—The death of the leopard— General plan of return.

THE history of an expedition divided into groups with distinct special aims, and busy simultaneously with their several labours in different places, is necessarily disconnected, and must now and again go back to take up another thread, and so follow the course of each separate section individually.

We must therefore beg the reader to return to the 22nd of June, when the Duke left Camp IV on the Scott Elliot Pass to descend into the Bujuku Valley and penetrate to the northern mountains. At this date Commander Cagni and Dr. Cavalli, and the guide Brocherel, were

Further Ascents and Work at Bujongolo.

ascending the Alexandra Peak in a dense fog. During the three hours they spent on the summit they had a few

CLIMBING THE ALEXANDRA PEAK.

glimpses of clear sky and were able to discern the neighbouring Margherita Peak and to repeat certain compass observations of the surrounding mountains. On their way back they had to wade through soft snow to the knee.

Vittorio Sella had left at daybreak with his photographic equipment and succeeded in getting a few views of the peaks from the ridges around the camp, while Roccati was collecting geological data and mineralogical specimens.

During the 23rd and the 24th the same storm which had rendered useless the Duke's first ascent to Vittorio Emanuele Peak and had kept him a prisoner in Camp V for two whole days, prevented Vittorio Sella and Roccati from accomplishing any sort of work outside of the tent.

Chapter IX.

As to Cagni, he was in a hurry to get back to Bujongolo as soon as might be, in order to lose no time in starting his magnetic observations and in calculating the formation of a base line, which was necessary to complete the triangulation. He left Camp IV on the 23rd with Dr. Cavalli, and the very same evening crossed the Freshfield Pass, and reached Bujongolo under pouring rain. He left deposits of rations along the way for the use of those who had remained behind. Dr. Cavalli remained at Camp III, at the foot of the western slope of Mt. Baker, to collect botanical specimens, and only reached Bujongolo on the following day, also in a completely soaked condition.

He found Cagni busy with all sorts of occupations. He had been working at organization, paying porters, etc., and was

MOEBIUS PEAK FROM THE SOUTH-EAST RIDGE OF THE ALEXANDRA PEAK.

now engaged in sending off small parties of natives to provide the Duke's party with rations in the far valleys to the west of

246

Further Ascents and Work at Bujongolo.

Mt. Speke. Several Bakonjo had bruised their feet and stood in need of the doctor's care. Profiting by the absence of the greater part of the tents, they proceeded with the work of improving the camp, enlarging the platforms already existing and forming new ones, filling up holes, moving blocks of rock and cutting down trees to increase the level space at their disposal.

The fearful weather prevented them from taking any observations. During a whole week Commander Cagni was not able to see the sun for a single continuous hour. The rainfall was slight but almost incessant, and the fog was so dense as to make it impossible to see the further side of the valley.

In spite of all this, Commander Cagni was able to take a few astronomical observations during fugitive moments of clear weather on the 25th, 27th, and 28th of June.

On the 25th, Vittorio Sella, taking advantage of a slight improvement in the weather, started from Camp IV with Roccati, Brocherel, and Botta and accomplished the ascent of the Moebius Peak, the only one of Mt. Stanley which had not yet been ascended. He then made a short excursion on the serpentine rocks of the western slopes, crossing the ridge after demolishing the great snowy cornice with the ice-axes. Here they had a view of two good-sized lakes in the valley to the west. They came back to camp under a heavy snowfall, but the day had not been wasted.

He set forth again on the following morning, by daybreak, with Brocherel and Botta. From the ice plain they saw the Duke on the summit of the Vittorio Emanuele Peak. They took photographs between one drift of mist and another, and in due time reached the summit of Alexandra Peak. The snow began to fall again as they returned to camp.

Chapter IX.

The 27th was an even more successful day for Vittorio Sella, who, accompanied by Roccati, first re-ascended Alexandra

SAVOIA, ELENA AND MOEBIUS PEAKS, AND MT. LUIGI DI SAVOIA SEEN FROM THE SOUTH-EAST RIDGE OF THE ALEXANDRA PEAK.

Peak, which was thus climbed for the fifth time, then returned to the Stanley plateau, and with Brocherel and Botta crossed the col between Alexandra and Moebius Peaks, and went about 1,300 feet down the broken western glacier. From a rocky spur projecting between the glaciers which descend from Moebius Peak and those which descend from Alexandra Peak he was able to take several photographs of the western slopes, thus getting a complete series of views of Mt. Stanley from every side.*

* The rocky spur at the foot of the western glaciers of Mt. Stanley, which was climbed by Vittorio Sella, comes out quite clearly in Stuhlmann's plate, reproduced on p. 206. The photographs taken by Vittorio Sella on this occasion are those which have enabled us to identify with Mt. Stanley the mountain represented in the above-mentioned plate.

Further Ascents and Work at Bujongolo.

Thence he re-ascended to the ridge and to the plateau, and returned with Roccati to the camp.

On the following day, in a storm of snow and hail, they struck camp with the assistance of the porters newly arrived from Bujongolo, and descended to the lakes to the west of Mt. Baker. On the 29th they again set up their tent on the Freshfield Pass. On the very same day the photographic camera was planted high on the south ridge of the Edward Peak, near to the edge of the glacier. After three hours of vain waiting under rain and sleet, they finally came down to the tent, leaving the camera where it stood. The

FRESH SNOW ON FRESHFIELD'S PASS.

whole of the following day was spent upon the ridge, crouching under the snowfall close to the camera. Even on the pass so much snow had fallen that it had brought down the tent. The firewood was soaked through and through, and in spite of copious libations of petroleum it was extremely difficult to kindle.

Chapter IX.

By the 1st of July, Roccati had finished his collection of minerals and rocks around this pass and the neighbouring glaciers. He therefore descended to Bujongolo, leaving Vittorio Sella alone with Brocherel and Botta, obstinately determined not to give up the struggle. In the afternoon the Duke also crossed the pass, returning from the far distant Mt. Emin and proceeding directly to Bujongolo.

On the following morning, in most unpromising weather, Vittorio Sella, with the two guides, climbed the Edward Peak directly from the col by the southern ridge. He was able to take an occasional photograph and an incomplete panorama. On the way down he was overtaken by a violent recrudescence of the storm, which lasted the whole of the next day with alternate snow and hail.

The spectacle presented by storms at that altitude (above 14,000 feet) is surpassingly grand. Heavy cumulus clouds

MT. STANLEY FROM FRESHFIELD'S COL.

hang over the Semliki River, which winds far off in the valley like a streak of silver. Huge bodies of whirling vapours rise

250

Further Ascents and Work at Bujongolo.

from the eastern and western valleys and strike one another with an incessant explosion of lightning and thunder, dissolving only to be replaced by fresh supplies from below.

Often of an evening after a day of fog, rain, snow and hail, the sky clears up. Through the moist atmosphere, as transparent as glass, the sinking sun appears like a vast globe of fire, suffusing the valleys, glaciers and snows to the westward with vivid flame colour.

On the morning of the 4th of July, Vittorio Sella with his two companions again left the tent to climb to the central

Sella Peak.   Weismann Peak.

MT. LUIGI DI SAVOIA FROM FRESHFIELD'S COL.

peak of Mount Luigi di Savoia which now bears his name. Crossing the head of the valley to the west of the Freshfield

Chapter IX.

Pass he reached a depression of the ridge. It was extremely difficult to find the way in the mist. Numerous aiguilles of rock obliged them to cross a steep névé to the south of the ridge and then to return to the north side under the summit, which they reached by a rocky gully. The Sella Peak, 15,286 feet, is rocky and dotted with numerous fulgurites. The edges of the slabs are here and there perforated to a depth of some inches and look as if they were worm-eaten. They

THE SOUTH RIDGE OF EDWARD PEAK AND THE CAMP CLOSE TO FRESHFIELD'S PASS.

spent several hours upon the summit without the chance of taking a single photograph. They were scarcely able, during a

Further Ascents and Work at Bujongolo.

momentary clearing of the mist, to distinguish the Weismann Peak to the south-west at the end of a long snowy ridge. On their return they descended straight to the bottom of the valley, which was full of watery and muddy spots, with the usual vegetation of senecio, and reached the tent after nightfall with fine moonlight.

Vittorio Sella finally rejoined the rest of the expedition at Bujongolo on the 5th of July, after a whole week spent upon the Freshfield Pass in fruitless expeditions up the ridges, and hours and hours of waiting beside his camera in the storms. For all his tenacity and energy he had not succeeded in getting a complete panorama from the Edward Peak as he had proposed to do.

The party at Bujongolo had not meantime remained idle. Commander Cagni had vainly attempted to take magnetic observations, but was prevented by the abundance of minerals containing iron in the rocks around Bujongolo. This influence was so considerable that it could be felt even when the inclinometer was placed at a height of some yards above the earth upon a wooden frame-work constructed for the purpose.

The greatest difficulty, however, was in finding a stretch of ground level enough and wide enough to allow of measuring a base line whose extremities were to be connected with two of the peaks forming a part of the network of angles measured by the Duke from the different mountains which he ascended.

There was a level place some distance back, above the cliff, at the foot of which stood the Camp of Bujongolo. But from this level space they could only see the Edward and Cagni Peaks which had not been connected with the others.

Chapter IX.

Another place higher up on the path leading to the Freshfield Pass, which the rains and the going to and fro of the porters had now reduced to the condition of a ditch full of mud, offered no better opportunities. The Duke and Cagni became convinced of this after spending a whole day there in the rain.

They accordingly planned to prepare a base line on the first-mentioned level behind Bujongolo. The Duke was then to re-ascend the Edward Peak, while Cagni was to climb the mountain which bears his name, and from these two they were to measure the angles of the other peaks. Everything now depended upon the good luck of getting a few hours of clear weather upon these two summits.

Meantime, on the 2nd of July, the Duke made a recognizance in the valley which runs between Mts. Baker and Cagni and comes out opposite Bujongolo. This valley he found to be barred by great steep slabs of rock, extremely slippery and certainly impassable for the native porters. The weather continued bad. Mt. Baker was completely covered with fresh snow. In the valley the rain had turned the whole ground into one mass of deep mud. On the 4th of July, between the showers, they succeeded, by taking advantage of every break in the fog, in tracing the base line upon the level tract above the camp, which consisted of a carpet of moss upon a muddy soil, dotted with senecios dripping with rain. In order to mount the theodolite at the extremities of the base line, they were obliged to build real foundations, sinking tree trunks into the mud more than six feet down to serve as piles.

Hardly had they taken these preparatory measures before the weather began to improve. On the 5th of July, on a

Further Ascents and Work at Bujongolo.

perfectly clear and very cold morning, the Duke again went up to the Freshfield Pass. Roccati, who had accompanied him so far, here re-descended to Bujongolo with Sella, while H.R.H. proceeded directly to the Edward Peak, following the southern crest along which Sella had made the ascent three days before. The mists returned before he reached the summit. It was only late in the afternoon that he was able to take a few angles in a brief moment of clear sky.

The Duke returned to the camp at nightfall. On the 6th of July the weather was again completely overcast and no work was possible, but on the 7th he returned early in the morning to the summit and was able to complete all the measurements.

On the following morning he ascended the Stairs Peak of Mt. Luigi di Savoia before returning to Bujongolo.

Commander Cagni in the meantime had left Bujongolo on the 6th with Joseph Petigax, Brocherel and a few natives to ascend the rocky peak to the north of the camp, which was to be connected on one hand with one extremity of the base line, and on the other with the net of angles of the different peaks.

The Cagni Peak, as may be seen upon the map, rises at the southern extremity of a buttress which runs between Mt. Baker and the South Portal Peak, flanked by two little valleys containing small lakes and tributary streams of the Mobuku.

Wishing to avoid the slabs of rock which had prevented the Duke in his recognizance of the 2nd of July from entering the valley to the west of the peak, and likewise to avoid crossing the Mobuku Valley below Bujongolo in the deep mire and through the dense heath forest, Cagni had decided to go

Chapter IX.

up the slopes of Mt. Baker and thence to traverse under Wollaston and Moore Peaks, towards the Cagni Peak.

Accordingly the party turned its steps first towards Grauer's Camp near to the Moore Glacier, and thence skirted the eastern slopes of Mt. Baker, intending to reach the col to the east of the Moore Peak. But their eternal enemy the fog obliged them to stop on the steep slope in the snow, stones and mud.

On the following day it became plain that it was impossible to pursue this route. It was necessary to go down to the narrow gorge between Wollaston Peak and Mt. Cagni. This was no easy task, and in more places than one they were obliged to let down the loads by a rope, and even to let down the porters as if they were parcels. Once at the foot of the south-west side of the Cagni Peak, which was quite perpendicular to the very bottom of the little valley, they ascended this latter as far as its head, through a dense wood of heath, and set up their tents immediately under the col.

From this point, on July 8th, they followed the spur which bears the Cagni Peak at its end along its whole length from north to south, keeping upon its western slope. In this way they reached the terminal cone, where they left their equipment, and after a short climb in the mist, about 3.30 p.m. they reached a small platform, which they took for the summit. The camp theodolite was at once set up upon its tripod. Suddenly through the mist they perceived to the south the real peak, which the refraction of the mist caused them to see as if at a very great height over their heads. The theodolite was immediately taken down, they descended from the little point which they had reached, and after a real Alpine climb up a very narrow ridge over a difficult bit of *arête*, about 12 feet high with

MOUNT CAGNI

Further Ascents and Work at Bujongolo.

insufficient handholds, and skirting round rocky *gendarmes* on their smooth, steep sides, they reached the real summit about six in the evening.

The mist had entirely disappeared, but nightfall was very near. Commander Cagni had scarcely time to take observations of all the peaks with the compass. They came down in the dark.

On the following morning by sunrise, the weather being perfectly clear, Cagni was once more on the summit, and was able to take measurements of all the angles with the theodolite and with the compass. They set up a stone man, and by eight o'clock they were preparing to return when the first mists began to rise. They came back by the same way, along the spur to the north of the peak and then down into the little valley to the west of it, which they now descended to the point where it opens into the Mobuku Valley. Here the mist, which had become dense, was added to all the other difficulties of crossing the tangled forest, which was very similar to the one above Kichuchu. They reached Bujongolo the same evening.

Sella was there alone, waiting for Cagni's Alpine tent to set forth upon a new photographic expedition. The Duke had gone up to Camp I upon Mt. Baker that very day. From this point on the following day, July 10th, through a gully to the east and then along the south ridge, he reached the Wollaston Peak, 15,286 feet, which had not as yet been ascended by any member of the expedition. The rocks were covered with ice. The weather was clear, and he was able to take observations for two whole hours. Next, following the high ridge, he traversed to the Moore Peak, whence he came down along the ridge which had already been climbed by Vittorio Sella, to the Grauer Col, and so back to Bujongolo.

257

Chapter IX.

Vittorio Sella had set forth in the morning with Botta and a few natives, and had returned to the Freshfield Pass. He did not return again to Bujongolo. On the 11th of July he was again upon the Edward Peak at sunrise, and was at last enabled to take the complete panorama of the chain for which he had once waited a whole week in vain on the Freshfield Pass. On the way back he paid a visit to the little knob somewhat lower down, which had been climbed twice by Wollaston, whose card he now found with the following inscriptions : " A. F. R. Wollaston, R. B. Woosnam, 17th February, 1906. Height by aneroid 16,050 feet."

" A. F. R. Wollaston (Alpine Club), R. B. Woosnam, D. Carruthers of the British Museum Expedition to Ruwenzori. Five hours from Bujongolo. Water boil. 183·6 ; temp. of air 39·7 ; aneroid 16,150 feet, 3rd April, 1906."

On the 12th of July, the weather remaining fair, Sella again ascended the Stairs Peak, where he took some good photographs.

In the meantime Commander Cagni had done two days' work in finishing the mensuration of the base line and connecting it with Edward and Cagni Peaks, and was able to complete an occultation, fixing the longitude and the latitude of one of its extremities.

In order to follow the intense activity of all the different members of the expedition occupied in such various ways and yet directed to one common aim, our story has necessarily become little more than a simple list of facts and of dates.

After the return of the expedition to Bujongolo, the leopard had resumed his daring visits to the camp, killing sheep and coming close to the fires among the native porters to steal the meat. Everyone was too busy to heed him. But the cook, Igini, with Bulli, planned an ambush with two rifles

Further Ascents and Work at Bujongolo.

and a piece of meat. One night the splendid animal fell into this trap and was killed on the spot with two balls through its skull.

On the 12th of July, the Prince was able to consider the work of the expedition as ended. On the 7th, Roccati, who had again returned to the Mobuku Glacier to put marks of red paint on the rocks at the limit where the ice stopped, and who had finished arranging all his collections, had already left Bujongolo with Cavalli and with a party of Bakonjo porters carrying a portion of the equipment, bound for Ibanda, the lowest camp in the Mobuku Valley.

One mountain alone remained unclimbed, namely, Mt. Gessi, and the Duke was not in a mood to leave it unattempted, all the more so as this ascent would be connected with an exploration of the Bujuku Valley as yet absolutely unknown and worth traversing in its whole length. A party of Bakonjo had started from the point where the Bujuku Valley opens into the Mobuku Valley opposite Nakitawa, and had already cut a rough track as far as the head of the Valley.

The plan was now for the Duke to descend the Bujuku Valley with Sella, while Cagni was to direct the transport of all the portion of the equipment which was still at Bujongolo down the Mobuku Valley, and was then to meet Cavalli and Roccati at Ibanda and there wait for the Duke. Thus Ibanda became the general rendezvous for the whole expedition.

CHAPTER X.

EXPLORATION OF THE BUJUKU VALLEY AND OF MT. GESSI.
RETURN OF THE EXPEDITION.

Departure from Bujongolo—The Camp of Ibanda—Visit to the Glacier Lake of the Mahoma Valley—H.R.H. leaves Bujongolo—Lake Bujuku—Descent of the Bujuku Valley—The Migusi Valley—Ascents of the Peaks of Mount Gessi— The Lower Bujuku Valley—The Expedition meets at Ibanda—Résumé of the Mountaineering Work of the Expedition—Return to Fort Portal—The Volcanic Region of Toro—Some Shooting—Arrival at Entebbe—The Ripon Falls— Departure from Africa—Sir Henry Stanley's wish realized.

ON the 1st of July, immediately after the Duke returned to Bujongolo from Mt. Emin, preparations had been commenced for leaving the mountains definitely, and had been carried on during the intervals left free by the varied work of the camp. In this way were gradually prepared the loads containing the scientific collections, the instruments, and all those objects which were no longer necessary.

They had at their disposal some forty Bakonjo porters only, as it would have been difficult to have provided supplies for a greater number so far from Ibanda. It was therefore necessary

Bujuku Valley.—Return of the Expedition.

to carry down the equipment in several trips. The first party of porters had left Bujongolo on the 4th July with forty loads. On the 7th a second caravan went down, accompanied by Roccati and by Cavalli, who had hastened his departure when he heard that there were porters ill in various camps of the valley, a report which proved to be without foundation. A week later Cagni left Bujongolo with Laurent Petigax, Brocherel, Igini, and twenty-three natives; the Duke had left for the Bujuku Valley on the previous day. Finally, on the 15th of July, the departure of Bulli with a last party of thirty Bakonjos left Bujongolo deserted.

All were satisfied with the work done, and were in fine spirits at the prospect of returning home, and left without a regret the wild rock which had offered them shelter during five weeks. They were glad to leave behind them so much mud and stones, the melancholy vegetation consumed by the mildews and lichens, the pallid light of the mists, the everlasting drip of the rain, the damp and the cold, and to get back to the sun and the dry heat of the tropical plains, the life and the colour, the cries of birds, the bright flowers and the gay crowd of thoughtless and noisy Bagandas.

The Mobuku River, swollen by more than fifteen days of continuous rains, was no longer recognizable. It formed magnificent cascades from one of the valley terraces to another. At every step on their way down, the parties met porters on their way up to Bujongolo to fetch loads.

A month before, when they first came up from the plain, the valley had struck them as almost without sound of animal life, but now, after weeks spent in the silence of the mountains where at the utmost an occasional crow hovered overhead, they were impressed by every buzzing of insects or fluttering of wings. Bihunga had become an important centre. It was the place where the Bakonjo porters passed on the loads to the Baganda of the plain who had again been concentrated at Ibanda.

Ibanda had now become a big permanent camp. The members of the expedition, as they arrived from Bujongolo, were received by the neighbouring chieftains with the usual ceremony and offering of gifts. The camp had become the natural meeting place for all the inhabitants of the surrounding

PORTERS IN CAMP, AT IBANDA.

villages. Women and old men were busy rooting out weeds and preparing the ground to construct new huts near the tents. From morning till night there was a racket and bustle; they all crowded around the fires, around the kitchens, around the barbers, while the native soldiers wandered hither and thither attempting to keep a little order in the confusion. The river was generally full of natives, bathing and disporting themselves in the water with great enjoyment.

Bujuku Valley.—Return of the Expedition.

Dr. Cavalli found plenty of occupation, and was busy for several hours every day with the sick people who came from far and near, supported or carried with loving care by their relations or friends. While undergoing trifling operations they would scream and weep, and immediately after laugh like children.

Commander Cagni undertook a series of magnetic observations. Dr. Roccati made geological and mineralogical excursions in the neighbourhood. One of these took him to the little lake above Nakitawa where the Mahoma Valley opens into the Mobuku Valley. This lakelet had been observed by Moore, Johnston, Dawe, etc. Mr. Freshfield calls it Lake Kobokora, but from all accounts it would seem that no one had yet actually reached it. To arrive thither from Nakitawa, Roccati had to pass through virgin forest without any track, while the Bakonjos cut the way with axes through bamboos, lianas and heaths. There were moraine ridges to cross, through an undergrowth so dense that in many places they actually walked upon the thickets, on an elastic cushion of branches and twigs several feet deep. Now and again, one of the Bakonjo guides would climb a tree to get his bearings.

Near to a fallen trunk they found traces of an old camp fire, a bit of newspaper and a sardine tin, no doubtful sign of the passage of a preceding explorer, possibly Dr. Wollaston or some other member of the British Museum Expedition. From this point they reached the lakelet in one hour. It is plainly a glacial lake, oval in shape, and running from south-east to north-west, with steep shores and surrounded by a narrow strip of mud, beyond which the deep water begins at once. There was fog round about, and complete silence, with no sign of animal life. Dr. Roccati collected plants and zoological

Chapter X.

specimens from the mud on the banks. Laurent Petigax and Brocherel returned later to the lake and were able to confirm the observation that it has normally no emissaries.

While the members of the expedition were thus occupied at Ibanda, the Duke of the Abruzzi was completing the exploration of the mountains. He had left Bujongolo on the morning of the 13th of July with the guides Joseph Petigax, Ollier, a native soldier, a boy, and seventeen native porters including the guide, a fine old man of fifty years. At the Freshfield Pass he was joined by Sella and Botta, and they proceeded together as far as Camp III at the foot of the western slopes of Mt. Baker.

The valley of the lakes, which they had so often traversed in rain and fog, now, on this fine clear day, seemed to offer an entirely new prospect. The sun, however, seems almost to strike a false note in the dense and melancholy forest of senecios. The helichrysums seem like skeleton flowers, and the scene is grim, sad, lifeless and brooded over by an oppressive silence.

On the following day, after a clear sunrise, the air again grew dark with mists. They climbed to the Scott Elliot Pass by the well-known way and set forth down along the gully towards the Bujuku Valley. Those who went ahead were in incessant danger of being hit by the stones which the numerous party of natives kept rolling down, in spite of all precautions.

From the foot of the gully, in a very short space of time, after crossing the grotesque forest of senecio mingled with clumps of everlasting flowers, and interrupted at one point by a brief marshy tract covered with reeds, they reached the shores of Lake Bujuku (12,855 feet), a splendid sheet of calm water upon which they saw a few duck. The view of the peaks of Mt. Stanley and Mt. Baker towering above them with their grim precipices was, beyond all comparison,

LAKE BUJUKU AND MOUNT STANLEY

Bujuku Valley.—Return of the Expedition.

grander than the mountain scenery at the head of the Mobuku Valley.

They soon discovered, among the mosses and reeds on the shore of the lake, the track prepared for them by the Bakonjo natives across the gently sloping plain of the upper valley. This plain ends in a gorge formed by spurs which run down from the Moore Peak of Mt. Baker and the Johnston

THE BUJUKU VALLEY.

Peak of Mt. Speke. Here there is a short, steep barrier, similar in every respect to those which intersect the Mobuku Valley. They now had a sight of the first heaths (12,297 feet), mingled with a few lobelias, which were nearly all dead.

Making their way down, now on the right hand and now on the left of the torrent, they reached a second plain, after

Chapter X.

which the valley again narrows into a gorge formed by the north-ward prolongation of the spur on whose southern extremity rises the Cagni Peak. This spur runs so far across the valley as almost to meet the long and considerable buttress which stretches from Mt. Speke eastward and forms so far the northern or left wall of the valley. Upon the ridge of this spur of Mt. Speke stands the extraordinary monolith of rock, with regular and architectural lines, which had been one of the first features noticed by them in the ascents of Mt. Baker.

The way leads down the right side of the gorge, which is clothed at the bottom with a dense forest of heaths, which would have caused them to waste a good deal of time had a track not been already cut. They came out of this gorge upon a third plain of more ample dimensions, into which open several tributary valleys from the north. One of these runs up to the north-west behind the east spur of Mt. Speke, and at its head forms the narrow gorge between Mt. Emin and Mt. Gessi. This is the Migusi Valley. Two more valleys, divided by a minor ridge, are traversed by the torrent Kurungu, which springs from a little lake fed by the Iolanda Glacier of Mt. Gessi, and the Waigga which also flows from a lake at the foot of the North Portal.

On reaching this plain, they encamped in a suitable place (11,503 feet), near a sheltering rock at the foot of a spur on the right side of the valley in a clearing of the forest all full of blossoming helichrysum. The spot was lovely, the slopes of the valley clad with dense forest, while before them towered up the rocky peaks of the North Portal.

This Camp, marked No. IX on the map, was their starting point for the ascent of Mt. Gessi, the last mountain still left unclimbed.

Bujuku Valley.—Return of the Expedition.

On the morning of the 15th, the Duke, with two guides and a few native porters crossed the plain to the north, steering towards a depression on the ridge to the north of the valley which runs down from Mt. Speke. This depression he reached

CAMP IX, BUJUKU VALLEY.

by ascending up a small lateral valley skirting the side in order to avoid the dense brush. From the ridge they went down into the Migusi Valley and followed for some distance the tracks of a leopard, which had made its way through the thickets.

267

Chapter X.

The Migusi Valley is also formed of a series of successive terraces. They ascended first one rise and then another, and finally reached a slightly inclined plain leading to the head of the valley where the narrow gorge between Mts. Emin and Gessi begins. They skirted the plain and ascended the right slope of the valley to a point not far from the end of the Iolanda Glacier. All of the Bakonjo were marching remarkably well. The hardest work was for the guides, who had to cut a path through the dense thickets of brush.

Camp X (13,668 feet) was set up close to the ancient moraine, only a few hundred yards from the present face of the glacier, which ends in broken séracs on the brow of a cliff. The senecios and helichrysums climb up a little higher than the point where the camp was fixed. The view from this high level over the great amphitheatre of mountains is one of the finest panoramas of the whole Ruwenzori range.

On the morning of the 16th there was hard frost all around the camp. The start was made before daybreak. First they ascended a gully overhung by the terminal séracs of the Iolanda Glacier. Then they crossed the rocks to the right of the gully and reached the snow, and then the south-east ridge of the mountain. At 6.30 a.m., the Duke set foot upon the rocky summit of the Iolanda Peak (15,647 feet). The rope had not been used in the ascent. Ollier began at once to build a monumental stone man.

The weather had been threatening when they set forth, but had now become quite clear, and the view of the mountains was complete in every detail, so that the Duke was able to make one more photographic panorama of the entire range. In this way the whole chain was photographed in panoramas taken from

268

THE BUJUKU TORRENT.

BUJUKU VALLEY

Bujuku Valley.—Return of the Expedition.

He set out with Botta on the morning of the 15th, and coming back to the second terrace of the valley proceeded to ascend the spur to the north, among giant heaths and shrubs of everlasting flower, over extremely broken ground, skirting

GIANT TREE IN THE LOWER BUJUKU VALLEY.

huge blocks, climbing upon fallen tree trunks covered with moss and concealing deep holes. The fog surrounded them before they had reached a point sufficiently high to get a clear view of the monolith.

273

Chapter X.

On the following day they were able to approach much nearer. Here a disappointment awaited Sella, for the monolith proved to be a very commonplace pinnacle of rock which had received from its isolated position on a ridge an appearance of being much more grand than was actually the case. He came back to Camp IX by night, a few hours later than the Duke.

On the 17th they proceeded to descend the valley under a clouded sky but without rain or fog. They crossed the

IBANDA.

terrace which forms the meeting point of the Bujuku, Migusi and Kurungu Valleys. This is probably an ancient lake bottom, and is now completely covered with reeds. They skirted its left slope on uneven ground scattered with little grassy hillocks. They now reached the third rise followed by a long gorge running between the two South Portals.

Bujuku Valley.—Return of the Expedition.

Here the valley bends slightly southward and the descent becomes steeper. They followed the left side of the river, which falls in rapids and noisy cataracts. A little below the plain the senecios come to an end, but the lobelias continue (Stuhlmanni and Deckenni).

After crossing the Manureggio, which is a tributary of the Bujuku and flows into it from the left, they entered a region of

BAGANDA VILLAGE.

giant grass which grows like a bed of reeds between the heaths, and covers on every side the steep declivities dotted with huge, boulders and intersected by a mass of little irregular gorges.

The natives had made a path by simply trampling down the long thick stalks, which form an elastic surface where you slip, sink in, and stumble at every step.

Camp XI (9,547 feet) was placed below the gorge of the Portals. They now found themselves once more in the true

Chapter X.

forest among huge trees, fine podocarpus, entwined with lianas and bamboo thickets. There were no more senecio nor lobelia. Troops of monkeys disported themselves in the branches, and the air was full of the song of birds.

In the evening it began to rain for the first time after ten days of dry weather. It was the only considerable period of

CRATER LAKE KAITABAROGA, NEAR FORT PORTAL.

good weather that the expedition had met with among the mountains.

On the following day they descended by a path which grew better and better, keeping upon the left bank of the stream. On reaching the bottom of the valley they turned southward, traversing the Bujuku and a small affluent, and made straight for the Mobuku where they found a rough bridge of bamboos.

Bujuku Valley.—Return of the Expedition.

Soon after they reached the familiar track down the Mobuku Valley and climbed the moraine of Nakitawa. Two hours more brought them to Bihunga and two more to Ibanda, where the whole expedition was now assembled.

The Duke of the Abruzzi had now fulfilled the task which he had undertaken; his untiring energy, seconded by the zeal

BACK AGAIN ON THE SHORES OF LAKE VICTORIA.

and ability of his companions, had brought the exploration of Ruwenzori to completion.

I have put together in tabular form a list of all the ascents of the Ruwenzori Peaks made by the members of the Italian expedition between the 10th of June and the 16th of July. This table shows the mountaineering work done by the expedition.

277

Chapter X.

Mountain.	Peak.	Height above sea-level.	Date.	By whom climbed.	Route followed.
Stanley	Margherita	16,815	June 18	H.R.H. the Duke of the Abruzzi Guides: J. Petigax, Ollier, and Brocherel	From the col between Alexandra and Margherita Peaks.
	Alexandra ...	16,749	June 18	H.R.H. the Duke of the Abruzzi Guides: J. Petigax, Ollier, and Brocherel	By the Eastern Ridge.
			June 20	H.R.H. the Duke of the Abruzzi	,,
			June 22	U. Cagni and A. Cavalli	,
			June 26	V. Sella ...	,,
			June 27	V. Sella and A. Roccati	,,
	Elena ...	16,388	June 20	H.R.H. the Duke of the Abruzzi Guides: J. Petigax, Ollier, and Brocherel	By an Eastern Gully.
	Moebius ...	—	June 25	V. Sella and A. Roccati Guides: Brocherel and Botta	By the Eastern Ridge.
	Savoia ...	16,339	June 20	H.R.H. the Duke of the Abruzzi Guides: J. Petigax, Ollier and Brocherel	Traversed.

The mountains are given in order of height.   Guides are mentioned in first ascents only.

Bujuku Valley.—Return of the Expedition.

TABLE OF ASCENTS IN THE RUWENZORI RANGE MADE BY THE EXPEDITION OF H.R.H. THE DUKE OF THE ABRUZZI, IN THE MONTHS OF JUNE AND JULY, 1906—*continued*.

Mountain.	Peak.	Height above sea-level.	Date.	By whom climbed.	Route followed.
Speke ...	Vittorio Emanuele	16,080	June 23	H.R.H. the Duke of the Abruzzi Guides: J. Petigax and Ollier	From the West.
	Johnston ...	15,906	—	Not climbed.	
Baker ...	Edward ...	15,988	June 10	H.R.H. the Duke of the Abruzzi Guides: J. Petigax, Ollier, and Brocherel	From Grauer Col.
			July 2	V. Sella ...	From Freshfield Col.
			July 5	H.R.H. the Duke of the Abruzzi	,,
			July 7	H.R.H. the Duke of the Abruzzi	,,
			July 11	V. Sella ...	,,
	Semper ...	15,843	June 10	H.R.H. the Duke of the Abruzzi Guides: J. Petigax, Ollier, and Brocherel	From Grauer Col.
	Wollaston ...	15,286	July 10	H.R.H. the Duke of the Abruzzi Guides: Ollier and L. Petigax	By a Western Gully and the South Ridge.
	Moore ...	15,269	June 12	V. Sella ... Guides: Brocherel and Botta	From Grauer Col.
			July 10	H.R.H. the Duke of the Abruzzi	Traversed.

The mountains are given in order of height. Guides are mentioned in first ascents only.

Chapter X.

Mountain.	Peak.	Height above sea-level.	Date.	By whom climbed.	Route followed.
Emin ...	Umberto ...	15,797	June 28	H.R.H. the Duke of the Abruzzi Guides: J. Petigax, L. Petigax and Ollier	By the South-West Ridge.
	Kraepelin ...	15,752	—	Not climbed.	
Gessi ...	Iolanda ...	15,647	July 16	H.R.H. the Duke of the Abruzzi Guides: J. Petigax and Ollier	Traversed.
	Bottego ...	15,483	July 16	H.R.H. the Duke of the Abruzzi Guides: J. Petigax and Ollier	By the South Ridge.
Luigi di Savoia	Weismann...	15,299	—	Not climbed.	
	Sella ...	15,286	July 4	V. Sella ... Guides: Brocherel and Botta	By a W. Gully and the North Ridge.
	Stairs ...	15,059	June 19	V. Sella and A. Roccati, without guides	By the Western Ridge.
			July 8	H.R.H. the Duke of the Abruzzi	,,
			July 12	V. Sella ...	,,
Cagni ...	...	14,826	July 8	U. Cagni ... Guides: J. Petigax and Brocherel	By the Northern Ridge.
			July 9	U. Cagni ...	,,

The mountains are given in order of height. Guides are mentioned in first ascents only.

Bujuku Valley.—Return of the Expedition.

It took the expedition two days to reach Fort Portal from Ibanda. Ruwenzori was again enveloped in its usual impenetrable veil of clouds and mists and they saw it no more. They were now again in the suffocating heat of the plain, among the noisy crowd of porters and the familiar scenes of native villages with their plantain groves, and again received at each stage by the chieftains with all the ceremonial of African etiquette.

At Fort Portal the English officials, King Kasagama with his court, and the missionaries rivalled one another in hospitality and courtesy toward H.R.H. and his companions.

While they were here, Roccati went with Sella upon a geological and photographic excursion to the craters and the crater lakes of the volcanic region of Toro. The shores of these lakes are covered with a dense vegetation of palms, dracenas, and euphorbia, which are mirrored in the water, while the water itself, the air and the wood swarm with an incredibly rich animal life, protected, perhaps, by the superstition which causes the natives to shun these craters as haunts of wizards and of evil spirits.

In the meantime, the Duke with Cagni and Cavalli, joined later by Sella, made some shooting excursions.

It was now the dry and less favourable season. It was impossible to penetrate the dense grasses which formed walls on either side of the paths and hid the surrounding country. Every night round Fort Portal the district was lit up with the red glare of the fires, which burned miles and miles of dry grass.

In the beginning of August the time came for their departure. Messrs. Knowles and Haldane accompanied them from Fort Portal. Notwithstanding the frequent storms,

Chapter X.

there were field fires in every direction, which even became a source of danger to the camps when the wind blew that way. Whole districts were quite bare and covered with ashes.

On the 7th of August, the expedition crossed the frontier between Toro and Uganda. Here it was met by Major Wyndham. The native porters seemed very impatient to get home and marched fast with few halts. The stages were differently distributed and the camps were set up in places where they had not stopped on the journey up.

A halt was made at Byndia, as previously at Kichiomi in the Kingdom of Toro, for the purpose of making a series of magnetic observations.

On the 14th of August, they at last reached the shores of Lake Victoria. The discipline of the caravan had become somewhat relaxed during the last days, and at every moment there were rows, disputes, and quarrels between the porters.

They had chosen a more direct route to return, and reached the banks of the lake just opposite the extreme end of the peninsula upon which Entebbe is situated. Here native canoes were ready in sufficient numbers to carry the whole party.

While the boats were being prepared and loaded, they lunched upon the bank of the lake in the shade of lofty trees. They reached Entebbe in the early afternoon.

After a week spent in packing the luggage which was to be carried back to Italy, and during which time they were entertained with the greatest hospitality and kindness by all the European residents, the expedition left Entebbe, with its crowd of islands and its flowery shores, upon the steamer *Sibyl*.

They stopped at Jinja to visit the famous Ripon Falls, which

BIPON FALLS.

form the origin of the Victoria Nile. Here they made an excursion in long native canoes upon the dark green waters of the river swarming with birds. On the 24th they reached Port Florence, and left the same day by train.

On the 28th of August the Italian expedition left African soil upon the steamer *Natal* of the French Messageries.

Five months later, before a largely attended meeting of the Royal Geographical Society, honoured by the presence of H.M. King Edward VII., H.R.H. the Duke of the Abruzzi gave an account of his discoveries, demonstrating that he had

HEAD OF THE VICTORIA NILE.

brought about the realization of the wish expressed five years before by Sir Henry M. Stanley before the same Society : " The dear wish that some person devoted to his work, some lover of Alpine climbing, would take Ruwenzori in hand and make a thorough work of it, explore it from top to bottom, through all those enormous defiles and those deep gorges."

Alas ! the great explorer died two years and a-half ago, and never saw his wish fulfilled.

285

Addendum.

[This book was already in print when **Mr. A. F. R. Wollaston** brought out his valuable book ("From Ruwenzori to the Congo," London, John Murray, published in September, 1908), in which, among other matters, he describes his climbs in the Ruwenzori chain. I have already dealt with his interesting mountaineering work in Chapters I and VII, in my sketch of the history of the exploration of Ruwenzori before H.R.H. the Duke of the Abruzzi.]

The Appendices have been translated by Prof. A. H. KEANE,
LL.D., F.R.G.S.

APPENDIX A.

Dr. LUIGI HUGUES.

THE MOUNTAINS OF THE MOON OF PTOLEMY'S GEOGRAPHY AND THE RUWENZORI RANGE.

THE MOUNTAINS OF THE MOON OF PTOLEMY'S GEOGRAPHY AND THE RUWENZORI RANGE.

IN Claudius Ptolemy's Geography (Book IV, chap. 8) we read as follows: "At the southern latitude of 12° 30', and between the longitudes of 57° and 67°, there rises the Mountain of the Moon, whose snows feed the lakes, sources of the Nile."

As under the latitude specified by the Geographer there is no high land in equatorial Africa that is elevated enough to be described as snowy, and still less as rising above the line of perpetual snows, and as, moreover, a latitude lying so far to the south would place such a high land quite beyond the upper basin of the Nile, the suspicion is not without justification that several geographers have raised that the mention of the Mountain (or of the Mountains) of the Moon does not come directly from Ptolemy, but is an interpolation foisted into his Geography by some Arab writer. This view is held by Cooley, who, in his *Ptolemy and the Nile*, published in 1854, thus expresses himself: "Ptolemy is a very methodical writer, and divides his Geography into chapters, each describing some natural zone or region, and containing connected information. Had he known that the lakes of the Nile were filled from the snows of mountains further south, he would, in conformity with his general method, have included these ultimate sources in his account of the river. Now the Mountains of the Moon are not mentioned in the chapter which treats of the Nile ([1]), but in a separate and, as it were, supplementary chapter, containing matters avowedly obscure and little known, and even there they are mentioned not directly, but in an oblique manner, and with a very suspicious gloss." ([2])

NOTE.—The figures in brackets in the text refer to the notes printed at the end of this Appendix.

289

Appendix A.

Dr. Heinrich Kiepert also appears to accept the same view where he writes in his *Treatise on Ancient Geography:* " The expression ' Blue Mountains (Jibel gomr), given by the Arabs to those great mountain masses (Kenia, Kilimanjaro and others), seen only from afar, and indistinctly, has long been wrongly interpreted in the sense of ' Mountains of the Moon ' (Jibel-el-Qamar), and thus gave rise to the translation Σελήνης ὄρος which is given on Ptolemy's map, and to an error which the recent explorations in that region of Africa have banished from our maps and from our books." (³) " The strange name of Mountains of the Moon," says Prof. Alfred Kirchhoff, " is due probably to an interchange of two Arab terms or to the twofold meaning of one and the same term." (⁴) And, in fact, the Arab writer el-Nowairi, quoted by Masudi, asserts that Kamar (read Qamar) means both *moon* and *white*. And in this connection it will not be beside the question to note that Aristotle had already placed the sources of the Nile in a ' Silver Mountain ' (Ἀργύρεος ὄρος). (⁵) This Silver Mountain has a striking analogy with the White Mountain of the mediæval Arab writers, an analogy which suggests some important and sensible reflections to Vivien de Saint-Martin. (⁶)

If the mention of the Mountains of the Moon, or else of the White Mountains (?) is of Arab origin, which, besides the stated reasons, might also be shown to be probable from the fact that no allusion to that lofty range is made in the edition of Ptolemy's Geography issued by Donis in 1482, (⁷) the latitude 12° 30′ S. would have been inserted in the text to bring it into accord with the position assigned by the Alexandrian Geographer to the two lakes, sources of the Nile. And respecting these lakes, here is what we gather from the seventh chapter of Book IV :—

The western lake has latitude (south) 6° and longitude 57°; the eastern is at latitude (south) 7° and longitude 65°. The rivers issuing from these two lakes unite at north latitude 2° and under the 60th meridian, and they thus form the chief branch of the Nile, which at north latitude 2° and under the 61st meridian receives the River Astapus, emissary from Lake Coloe, which lies on the equinoxial line and under the 69th degree of longitude.

It is quite understood that the Ptolemaic data referring to geographical features are not to be taken literally. The number of astronomic observations at the command of Ptolemy was very limited ; the results of those few observations, especially for the longitudes, were nearly all very far from the actual. To accomplish the gigantic work that he had undertaken, no better means occurred to the Geographer than that of reducing to astronomic data the elements—distances and directions—derived from the itineraries both by land and water, or already known from previous works, amongst which, first and

The Ruwenzori Range.

foremost, was that of his immediate precursor, Marinus of Tyre, or else those gathered by himself from the more or less accurate reports of travellers and seafarers. ([8]) All can see how defective such a method must be. From the early itineraries traced without compass in determining the directions, without chronometers for the intervals of time and distances, and without sufficient knowledge of the marine and atmospheric currents, it was obviously impossible to obtain other than quite hypothetic, and for the most part only roughly approximate results. ([9]) The reduction of the route distances to astronomic notations (degrees and fractions of degrees) was made by Ptolemy with the stadium unit equivalent to the 500th part of the equatorial degree. ([10]) But we know that those routes were based on a different unit of measure, namely, the Olympic stadium of 600 to the equatorial degree. Hence, if for instance, it was a question of an itinerary of 3,000 stadia (in the direction of the meridian), the number of corresponding degrees would be 5° of latitude according to the Olympic measure, while according to Ptolemy it came to 6°. And in general, to obtain the true, or the approximately true, differences of latitudes and longitudes, we have to multiply by $\frac{5}{6}$ those given by the Geographer, or, which is the same thing, reduce them by $\frac{1}{6}$. At the same time this single operation is very far from sufficing to introduce any accuracy into the Ptolemaic tables. It cannot be asserted in the first place that all the itineraries without exception were recorded in Olympic stadia; nor is the possibility to be excluded that for some of them the stadium of Eratosthenes of 700 to the equatorial degree was taken as the unit; in which case the reduction should be by $\frac{2}{7}$. Moreover, in a great many cases there occur errors of another nature, amongst which outstanding are those derived from the imperfect knowledge possessed by the ancients of many places and countries, from the inevitable inaccuracies in the calculation of distances and in determining the relative positions, from the windings of the route followed and so on. Despite of all this it is remarkable, not to say absolutely astounding, that the above-mentioned single reduction by $\frac{1}{6}$ suffices for the geographical sketch of the Upper Nile lands, such as is drawn by Ptolemy's Geography, to correspond broadly if not precisely with that presented to us by the modern maps. On this no doubt quite casual coincidence it will not be useless to dwell for a moment.

The latitude of Alexandria is given by Ptolemy as 30° 30′ N. (it is really 31° 12′); from Alexandria to the parallel of the eastern lake are therefore reckoned 37° 30′ E. = 37° E. Now the $\frac{5}{6}$ of 37° 5′ are equivalent to 31° 25′ = 31° 15′, and that lake thus falls under 0° 45′ south latitude. A similar calculation for the western lake brings us to north latitude 0° 9′. ([11]) These

291

Appendix A.

latitudes are very closely those of the northern shores of Lakes Victoria Nyanza and Albert Edward.

I come now to the longitudes. That of the western lake (57°) differs scarcely 3° from the longitude of Alexandria (60° according to Ptolemy), hence actually only 2° 30′ by the above-mentioned reduction. We have, therefore, a result little inferior to the reality, since the longitudes of Alexandria and of the west side of Lake Albert Edward are relatively to the meridian of Greenwich 30° and 29° 30′ (both E.) respectively, according to Stanley's map. The longitude of the eastern lake is 65° in Ptolemy, as above stated. It would consequently lie to the east of the meridian of Alexandria, and at a distance of 5° (4° 10′) according to the reduction. Now the mean longitude of Lake Victoria is 33° 15′ E., so that the difference is only minus 0° 55′. Thus in respect of the longitudes also there is nothing to prevent the identification of the two Ptolemaic lakes with Lakes Albert Edward and Victoria.

The confluence of the two effluents is placed by Ptolemy under the meridian of Alexandria ([12]), and in the north latitude of 2°. Hence it may fairly be placed where the river called the Somerset Nile by Speke enters Lake Albert, from which it soon again issues. Its latitude is little more than 2° N., while its longitude does not greatly exceed 30° E. Everything might therefore be reconciled by accepting Ptolemy's figures without any serious modification. On the other hand, by the process of reduction we get for the point of confluence 6° 45′ north latitude. It is, however, to be noted that somewhere about this latitude the main stream of the Nile begins to traverse a marshy region watered by several rivers nearly parallel to it, amongst them the Bahr el-Zaraf, the Nam Rol, and others, and that further on, towards latitude 9° N., the Bahr el-Abiad (White Nile) is joined both by the Bahr el-Ghazal coming from the west, and the Sobat from the east. To me the hypothesis does not seem at all too daring that precisely in this region the Alexandrian Geographer placed the confluence of the two upper branches, on the mistaken assumption that one of those rivers trending north was in fact the emissary of the eastern lake, just as for some years after Speke's memorable expedition Lake Baringo was supposed to be a north-eastern feeder of Lake Victoria, and had for its emissary the Asua, which is now known to flow, not to the lake but straight to the Nile at Dufile. ([13])

The almost perfect agreement of the results of modern research with the Ptolemaic data regarding the geographical features of the two lakes, sources of the Nile, is, I repeat, to be considered as a mere coincidence. Still the idea entertained by the great geographer on the general disposition of the upper basin of the Egyptian river was, broadly speaking, correct. And this might

The Ruwenzori Range.

at first sight be explained by admitting that those notions about the hydrographic relations might have been gathered by Ptolemy with the help of itineraries made along the valley of the river itself and generally in the direction from north to south. ([14]) Only, as Ptolemy himself says, these particulars were extant, at least in part, in the work of Marinus of Tyre, who in his turn had derived them from one of the then recent reports of the first Greek navigators of Egypt, who frequented the markets of East Africa from Cape Aromata to Cape Rhaptum ([15]): "After this he (Marinus) says that in the voyage between the Aromata and Rhaptum promontories a certain Diogenes . . . was in the neighbourhood of the Aromata, driven by the northern winds, and having on his right hand the Troglodytica arrived in five days at the lakes where the Nile rises, these lakes being somewhat more to the north than Rhapta." ([16])

In this the geographer of Tyre is contradicted by Ptolemy, who a little further on says: "The lakes whence rises the Nile are not near the sea, but far more inland on the Continent." This is an important correction very probably suggested to Ptolemy by the reports of those Greek seafarers, since the places from time to time visited by them on the east coast of Africa were not only important from the commercial standpoint, but also as so many centres whither fresh and numerous particulars could not fail to come to hand about the geographical and natural conditions of the inland regions. No wonder, therefore, if amongst those particulars was also that most important one regarding the existences of two lakes; and as the emporium of Rhapta, a place of great consequence and spoken of by Ptolemy as a metropolis ('Ραπτά μητρόπολις), is placed by him under the latitude of 7° S. ([17]), while, on the other hand, he was naturally inclined to believe that the two lakes lay due west of Rhapta, or nearly so, he accordingly gave to the eastern lake the same latitude of 7° S. and to the western 6° S. And I may here remark that, the position assigned by Ptolemy to Rhapta being almost exact ([18]), we may consider this place as a second centre of observations, such as those above described as having been carried out for Alexandria. Now, according to the tables, the longitude of Rhapta is 71°, and that of the eastern lake is given as 65°, the difference (6 degrees) being with the reduction 5°, and that is the difference between the mean longitude of the mouth of the Pangani (39°) and that of Lake Victoria (33° 15′). The 14 degrees of longitude that extend from the western lake (longitude 57° according to Ptolemy) to Rhapta (71°) are reduced to 11° 40′, and this scarcely exceeds the real difference (39°–29° 30′) by 2° 10′.

The almost identical results at which we arrive by taking as centres of

Appendix A.

astronomic studies the city of Alexandria and the commercial emporium of Rhapta, lead of themselves to the hypothesis that, besides the details gathered from the Greek seafarers along the east coast of Africa, the Alexandrian Geographer also utilized the information obtained in the valley of the great river itself. Nor will this assumption appear too bold if we bear in mind that long before the time of Ptolemy, the great Eratosthenes, speaking of the chief branch of the Nile, expressed himself thus: "Two waterways fall into the Nile: they both come from certain lakes lying far to the east and enclosing a very large island known by the name of Meroe. One of these waterways, called Astaboras, forms the east side of the island, the other is called Astapus. Some authors, however, give to the latter the name of Astasobas, and apply the name Astapus to another waterway, which they bring from the lakes lying in the region of the south, and regard it in some way as the main stream or else as the chief branch of the Nile, adding that its periodical floods are due to the summer rains." ([19]) If, as seems beyond doubt, the Astaboras is to be identified with the Atbara, the Astasobas with the Bahr el-Azrek or Blue Nile, and the Astapus with the White Nile or Bahr el-Abiad (main branch of the Nile), Ptolemy would have but repeated on the whole what three centuries before him had been so well expressed by the Librarian of Alexandria, merely adding on his own part the positions—latitude and longitude—of the two lakes lying in the region of the south, that is, south of the equinoxial line whose waters feed the chief artery of the all-important fluvial system.

At this point we meet with an apparently serious objection. According to the authors alluded to by Eratosthenes the name Astapus is given to the principal branch of the Nile flowing from the south, whereas Ptolemy applies it to an affluent of the Nile which, issuing from Lake Coloe under the equinoxial line, falls into the main stream at latitude 12° N. But, as above stated, the objection is only apparent. Eratosthenes, speaking for himself, had already given the name of Astapus to the river marking the west (and south-west) side of the island (peninsula) of Meroe, that is to say, the Abai or Bahr el-Azrek. Ptolemy, on his part, did not think it right to depart from the opinion of his predecessor, and so kept the name of Astapus for the subordinate river. It might be more important to notice in the Geography of the Alexandrian two errors, one of which affects the Lake Coloe (Lake Tana or Tsana in the heart of Abyssinia), which he places under the equinoxial line. The other mistake consists in describing the Astaboras as a river which mingles its waters with the Astapus. But an enquiry into all these matters, besides requiring too great a development, would be foreign to the question of the lakes, sources of the Nile, and to that of the Mountains of the Moon. Notice can only be taken of the

The Ruwenzori Range.

mistake made by Stanley, who, after calling Ptolemy "the Ravenstein or the Justus Perthes of his period" (Vol. II, p. 270), says that the easternmost lake was called by Ptolemy *Coloe Palus*, whereas this lake is expressly described in the Alexandrian's work as belonging to the secondary basin of the Bahr el-Azrek. ([20])

Meanwhile, from the facts so far pointed out, we clearly see how greatly those authors are at fault who place the two lakes of the Upper Nile, and as a necessary consequence the Mountains of the Moon too, in the highland region of Abyssinia, thus turning the Ptolemaic data upside down, and stating in support of their assumption that the ancients knew of only one system of snowy mountains in Africa, namely, that of the Abyssinian Semen. All the less can we accept the opinion of those writers who, with Ravenstein ([21]), prefer Marinus Tyrus to Ptolemy, and locate the Upper Nile lakes in the neighbourhood of the east coast, and precisely in the territory of the Afars (Dankali), that is at over 11° of north latitude.

Those two famous lakes are, beyond doubt, identical, the eastern with Lake Victoria, the western with Lake Albert or Albert Edward, or probably with both of them taken together. Nor does the objection hold which is suggested by the too great difference (8 degrees) in the longitudes of the two lacustrine basins, as, besides the uncertainty in which Ptolemy leaves us regarding the locality of the eastern lake, of which he gives us the geographical relations, it may be remarked that the difference might have been caused by the windings of the routes that had to be traversed to get from the southern shores of the eastern lake to any point of the western. ([22]) It is further objected that Ptolemy tells us nothing as to the size of the two lakes, which seems strange, especially as regards Lake Victoria, a rival in area of the largest lakes in the Laurentian basin of North America. On this point I may remark that neither for any of the other lakes does Ptolemy tell us anything respecting their extent. Why, then, should he make a solitary exception in the case of the two Nilotic ones ? Nor should it be forgotten that in his comprehensive work Ptolemy shows himself more especially in the light of an astronomer. The geographer appears, so to say, only in the second place. In fact, no trace is to be found of a physical description of the world, of its morphology, or of any of the other subjects that form the main object of pure geography. In this respect Ptolemy is far inferior to Strabo. His chief aim, says Bunbury, was to rectify the general map of the habitable globe, not only by supplying what had remained unknown to his predecessors, but also by applying from beginning to end a more scientific system based on solid astronomic foundations. He again inclined to the idea that had long before been entertained by

Appendix A.

Hipparchus, but which that great astronomer had been unable to realize owing to the great lack of materials.

The identity being thus demonstrated of Ptolemy's two lakes with Lake Victoria and the Albert-Albert Edward basin, we now come to the question of the Mountains of the Moon. That these uplands, lofty enough to feed the reservoirs of the Upper Nile with their snows, are to be placed amid the Abyssinian highlands, and more particularly in the mountains of Semen or of Gojam, is a view that must be absolutely rejected. To it are opposed the northern position of those mountains, the fact that the Abyssinian heights stand altogether outside the upper basin of the White Nile, and perhaps also the altitude itself which, although in some of its peaks rivalling that of Monte Rosa, is still too low to justify Ptolemy's statement, if, indeed, the Geographer intended to speak of perennial snows.

The Abyssinian Mountains being excluded, only two upland groups remain that might aspire to the honour of being identified with the Mountains of the Moon. These are the Kilimanjaro-Kenia ([24]) and the Ruwenzori groups. There is, however, a serious objection. Ptolemy (or the Arab interpolation ?) expressly states that the longitudinal axis of those mountains is developed in the equatorial direction along the parallel of 12° 30′ S. The Kilimanjaro-Kenia group is, on the contrary, developed in a direction which departs little from the meridian, while that of Ruwenzori has a trend nearly from S.S.W. to N.N.E. Nevertheless, this at first sight most formidable objection may perhaps be overcome, or better, toned down with a consideration of general hydrography. We know that, in accordance with their direction relatively to that of the lines of water-parting, rivers are normally divided into transversal and longitudinal. The first have a trend perpendicular, or nearly so, to the main water-parting line, while the second are parallel, or nearly so, to the same line. To which of these two categories belongs the course of the Upper Nile ? If we regard as a line of water-parting the undulating ground traversing Unyamweziland, and enclosing on the south the Upper Nile basin, and more particularly that of Lake Victoria, then the Upper Nile will be classed amongst the transversal rivers. If, on the other hand, we take as main dividing lines either the steep eastern scarp of the great African tableland (the watershed towards the Indian Ocean), or else the series of heights separating Lake Albert Edward, the Semliki valley, and Lake Albert from the Congo basin, then the Upper Nile will become a longitudinal river. Now, bearing in mind the decided trend of the Nile in the meridian direction, it is highly probable that we should incline rather to the first than to the second view, and accordingly place the region where the river rises in an upland tract running perpendicularly to its trend

The Ruwenzori Range.

that is, from west to east. But at the same time, either owing to our knowledge of the existence of snowy peaks in that part of east equatorial Africa, or else because of the generally admitted principle that the larger rivers rise in the highest mountains, ([25]) those moderate undulations of the land were without more ado transformed to a group of gigantic highlands. We thus see how, despite their trend, quite different from the equatorial, the two groups of Kilimanjaro-Kenia and Ruwenzori, thanks to their great elevation, came to form part of the Mountains of the Moon. ([26]) To which of the two should the preference be given?

Respecting Kilimanjaro-Kenia, we have to consider a fact of vast geological and hydrographic importance. The narrow strip of seaboard along the Indian Ocean, where prevail the jurassic limestones and argillaceous schists, is followed westwards by a chain of isolated crystalline heights commonly designated by the name of the East African Schistose Mountains. West of this system we enter a zone highly remarkable for its great geological disturbances. It is distinguished above all by the great East African Rift Valley, a vast line of fissure running in the direction of the meridian, and extending for 40° of latitude from the Asphaltites Lake (Dead Sea) all the way to Ugogo. The trough on the east side is to be regarded as a secondary rift, above which rise Mounts Meru, Kilimanjaro, and probably also Kenia. The whole of this district west of the East African Schistose system sends none of its running waters either directly or indirectly to the Indian Ocean. In other words, it is essentially a landlocked continental region. ([27]) Thus, while the east slope of the Schistose Mountains is traversed by streams tributary to the Indian Ocean, the few rivers of the west slope find no other outlet but the chain of lakelets which follow in the direction from north to south along the meridian rift. The aforesaid Kilimanjaro-Kenia group stands therefore absolutely outside the Lake Victoria and Somerset Nile basins. ([28])

It is otherwise with Ruwenzori, which, by its east watershed not only belongs to the basin of the Somerset Nile and of the region north-east of Lake Albert Edward, but also, by its south and west slopes, to the basin of the same Lake Albert Edward, the Semliki and Lake Albert. Hence, if, as is probable, there exists any orographic, if not geological, link between Ruwenzori and the group of Virunga Mountains, which rise to the south and south-west of Lake Albert Edward to an altitude of 13,000 feet, the identification of this highland system with the Mountains of the Moon would be all the more confirmed. This system is, in fact, the only one in the whole of equatorial Africa that completely satisfies all the conditions specified in Ptolemy's Geography, not even altogether excepting that of the general trend, which is precisely

297

Appendix A.

equatorial in the group of the Nfumbiro (properly Muhawura—" far seen ")
and Virunga Mountains, although these are far less elevated than Ruwenzori. ([29])

NOTES.

([1]) This is the seventh chapter of Book IV. After indicating the geographical positions of the two lakes sources of the Nile, it would naturally occur to Ptolemy to give that of the more southern snowy mountains. But he does not do so, and only speaks of them incidentally in chapter 8 of the same book, where there is no further reference to the Egyptian river.

([2]) COOLEY, *op. cit.*, pp. 77–78.

([3]) H. KIEPERT, *Lehrbuch der alten Geographie*, p. 210, note 2.

([4]) *Geogr. Mit.* 1892, *Litteratur-Bericht*, No. 49.

([5]) *Meteor.* Book I, chap. 13.

([6]) *Histoire de la Géographie*, pp. 109 and 124; *Le Nord de L'Afrique dans l'Antiquité grecque et romaine*, pp. 21 and 486.

([7]) MALFATTI, *Scritti geografici ed etnografici*, p. 454.

([8]) See in this connection the important considerations developed by Ptolemy in chaps. 4 and 6 of Book I.

([9]) The defects of this method, and the serious errors committed by Ptolemy in recasting and expanding the work of Marinus of Tyre, are excellently exhibited, with his usual clearness and shrewdness, by VIVIEN DE SAINT-MARTIN (*Histoire de la Géographie*, pp. 200 and 201). On Ptolemy's geographic system specially valuable are the pages devoted to this subject by BUNBURY in his *History of Ancient Geography* (2nd edition, Vol. II, pp. 546–579).

([10]) Very numerous instances of such numerical reductions are found in the *Geography*, and especially in Book I.

([11]) $(30° 5' × 6°) × \frac{5}{6} = 30°, 415$; $30° 5' - 30°, 419 = 0°,085$.

([12]) *Ptol. Geogr.*, Book IV, chap. 7. Here is the confluence of the rivers that flow from the southern lakes; long. 60°, lat. 2° N.

([13]) Such is also the opinion of not a few modern geographers, amongst whom I have pleasure in mentioning Dr. FELIX BARLIOUX in one of his learned dissertations published in 1874, under the title, *Doctrina Ptolemaei ab injuria recentiorum vindicata*, p. 31.

The Ruwenzori Range.

[14] "His (Ptolemy's) latitudes and longitudes are clearly worthless, except in so far as the former represent the broad fact that these lakes, and therefore the sources of the Nile, were actually situated south of the equator." So BUNBURY in the quoted work, Vol. II, pp. 614-15.

[15] Cape Aromata is usually identified with Cape *Guardafui*. HENRY SCHLICHTER (*Proc. of the Royal Geographical Society*, 1891, p. 529), places it much farther south, and identifies it with *Ras Aswad* (lat. 4° 30′ N.). Cape Rhaptum is placed by Ptolemy at one and a-half degree from the commercial emporium of Rhapta in the direction of the south. Touching its identity with any of the coast headlands in that part of Africa, geographers are not quite of accord. Müller places it at *Ras Puna*, Berlioux and Schlichter at *Ras Mambamku*. Nor is it easy to indicate the position of the commercial emporium of Rhapta, since it did not lie on the coast, but somewhat inland. Still, as the River Rhaptus of Ptolemy's Geography is most probably identical with the Pangani, not a few geographers place Rhapta on the lower course of that river. Bunbury (*op. cit.*, p. 454), says that Rhapta stood at the head of the bay opposite Zanzibar, not far from Bagamoyo.

[16] *Geogr.*, Book I, chap. 9.

[17] Admitting that Rhapta corresponded to some place on the lower course of the Pangani, Ptolemy's latitude 7° S. would differ by 1° 30′ from the actual, the mouth of the Pangani being at 5° 30′. If we locate Rhapta with Bunbury in the neighbourhood of Bagamoyo, the agreement will be almost perfect. In any case, the nearly correct description of the eastern seaboard is easily explained when we remember that, as we know from the *Periplus Maris Erythraei* and from the language of Ptolemy himself, the coastlands north of Rhapta were at that time very well known.

[18] *See* the foregoing note.

[19] STRABO, *Geogr.*, Book XVII, chap. 1, 1 ; BERGER, *Die geographischen Fragmente des Eratosthenes*, Vol. I, p. 302 sq.

[20] STANLEY, *In Darkest Africa*, Vol. II, p. 270.

[21] *Proceedings of the Royal Geographical Society*, 1891, p. 550.

[22] H. SCHLICHTER in *Proceedings of the Royal Geographical Society*, 1891, p. 534.

[23] BUNBURY, *History of Ancient Geography*, Vol. II.

[24] We know that the first notions regarding these gigantic mountains of East Africa date from the travels of the missionaries Krapf and Rebmann (1848-1851).

[25] The *Montes Atrapei* of European Sarmatia may serve as an instance.

[26] Bunbury argues much to the same effect. " The precision with which he determines the position and limits of a range of mountains, concerning which he had no real knowledge, and which had no existence in fact, finds a parallel in that of the Hyperborean Mountains in European Sarmatia ; and there seems no doubt that the process by which Ptolemy arrived at his conclusion was much the same in both cases. In this instance he had learned the existence of two lakes, which he believed to be the sources of the Nile ; he had learnt also the existence of a range of mountains, *some of which were so lofty as to be covered with snow*, though situated under the equator; he then at once assumed that the lakes were fed by the snows of the mountains, and having no real idea of the position of these last, drew them on his map in a straight line, to the south of the lakes, extending far enough to the east and west, to supply, as he conceived, the necessary drainage." *See History of Ancient Geography*, Vol. II, p. 616. It is needless to observe that the learned historian does not admit with Cooley the

Appendix A.

interpolation of the passage in the Geography where allusion is made to the Mountains of the Moon, or, in other words, he holds them to have been written by Ptolemy himself. " The attempt of Mr. Cooley," he writes, " to discard altogether the Mountains of the Moon as an interpolation in the text of Ptolemy, due to the Arabian Geographers, appears to me wholly untenable. The passage in which he speaks of them (IV, 9, 3) is unconnected with that concerning the two lakes (IV, 8, 23), and probably derived from a different authority; but it is not inconsistent with it." (*See op cit.*, p. 617, note 3.)

(27) O. BAUMANN, *Durch Masailand zur Nilquelle*, p. 133.

(28) Before these geographical details were known, geographers were naturally inclined to identify those snowy mountains of East Africa with the Mountains of the Moon of Ptolemy's Geography. It will suffice to mention CHARLES BEKE (*On the Mountains forming the eastern side of the Nile*, Edinburgh, 1861); VIVIEN DE SAINT-MARTIN (*Le Nord de l'Afrique dans l'Antiquité grecque et romaine*, Paris, 1863); ÉTIENNE FELIX BERLIOUX (*Doctrina Ptolemaei ab injuria recentiorum vindicata*, Paris, 1874), Sir E. H. BUNBURY (*A History of Ancient Geography*, Vol. II, p. 617); H. TOZER, who, in his *History of Ancient Geography*, published in 1897, hence subsequently to Stanley's last great expedition, writes at p. 352 : " The intelligence which is contained in these two statements (regarding the two lakes as sources of the Nile and the Mountains of the Moon) was probably transmitted, not by way of the Nile Valley, which was not followed by traders beyond the marshy region which has been already noticed, but from the coast in the neighbourhood of Zanzibar, where the station of Rhapta had been established. On this supposition it is not improbable that the lakes here spoken of are the Victoria and Albert Nyanza, and the mention of so unusual a phenomenon as snow-covered mountains in the neighbourhood of the equator supports the conjecture that the Mountains of the Moon are none other than Mounts Kilimanjaro (19,700 feet), and Kenia (18,370 feet), which lie between those lakes and the sea."

(29) Amongst the most vigorous champions of Stanley's view is H. S. SCHLICHTER, who concludes his learned work on PTOLEMY's *Topography of Eastern Equatorial Africa* (1891), with the following words :—" Mr. Stanley's discovery of this great snow mountain, surrounded by a series of other peaks, forms, so to speak, the key to the whole question of the Mountains of the Moon. For it is perfectly clear that by the Ptolemaean mountain, the snows of which feed the Nile lakes, only Ruwenzori can be meant, as may be seen from a glance at Mr. Stanley's map, where we find a great number of rivers (I have counted more than forty) which flow from the heights of Ruwenzori into the Semliki or the Albert Edward Nyanza. We have seen that the western end of the Mountains of the Moon, as described by Ptolemy, coincides with Ruwenzori, and Mr. Stanley is therefore perfectly justified in claiming to have found and identified the lofty peaks, celebrated in antiquity, in which the Nile takes its rise, and which, for many centuries past, were more enigmatical than any other mountain in the world."

Dealing with a question whose final resolution, in the absence of safe and positive data and in the scarcity of actual facts, must always remain a " pious wish," one well understands how Schlichter's conclusions were not unanimously accepted, and even found formidable opponents, amongst whom Ravenstein must be specially mentioned. The examination of the arguments advanced for and against would far exceed the modest limits to which I have confined myself in these pages. I must rest satisfied with here quoting the opinion expressed

The Ruwenzori Range.

on the subject by SIR HENRY H. JOHNSTON in his recent work, *The Nile Quest*, p. 28 : " The present writer is unable to understand why that able geographer, Mr. E. G. Ravenstein, has doubted the identification of Ruwenzori with Ptolemy's Mountains of the Moon. It must be obvious, when all facts are considered, that Ruwenzori was the principal germ of this idea. The Greek traders at Rhapta (Pangani) no doubt had some idea of the existence of Kiliman- aro, but it is doubtful whether either the single dome of Kilimanjaro or the gleaming pinnacle of Kenia would impress the imagination so strongly as the whole brilliant range of Ruwenzori's four or five snow peaks and thirty miles of glaciation."

APPENDIX B.

ASTRONOMIC, GEODETIC AND METEORO-LOGICAL OBSERVATIONS.

I.—REPORT ON ASTRONOMIC OBSERVATIONS,

By P. CAMPIGLI.

II.—GEODETIC OBSERVATIONS,

By P. CAMPIGLI.

III.—REPORT ON METEOROLOGICAL AND ALTIMETRIC OBSERVATIONS,

By PROF. D. OMODEI.

IN this note are contained the relations and calculations of the astronomic, meteorological and geodetic observations which H.R.H. the Duke of the Abruzzi was able to carry out on the route from Entebbe to Bujongolo, and during the exploration of the Ruwenzori Range.

The calculations relating to these observations, as well as the construction and plan of the topographic maps accompanying the present volume were executed at the Hydrographic Institute of the Royal Navy at Genoa.

The way by which the astronomic and meteorological observations were made, from which were obtained the positions and altitudes of the various points indicated on the maps, as well as the methods of calculation employed, are all embodied in the accompanying special reports drawn up through the care of the Director of the said Institute, Mattia Giavotto, Captain of frigate, the sections dealing with the meteorology and the astronomic observations being prepared by Prof. Omodei and the " Capo-Tecnico " Sig. P. Campigli respectively.

I.—REPORT ON ASTRONOMIC OBSERVATIONS.

By P. Campigli.

The astronomic determinations made by H.R.H. the Duke of the Abruzzi on the route between Entebbe and Ruwenzori are the result of solar observations made with an aluminium sextant, which was constructed in the engineering workshop of the Naval Hydrographic Institute at Genoa. Its graduated arc has a radius of 145 mm. (about 6 inches), being so subdivided as to show the 20 seconds on the vernier. Magnaghi's astronomic circle was used only in the very few cases where, for observations at the meridian or in its neighbourhood, the height of the sun was such as to make the use of the sextant less convenient.

Of course, all measured heights were duplicated at an artificial mercurial horizon, care being taken to reverse the position of the roof at half of each series of observations, in order to lessen to the utmost the influence of errors in case the glasses of the said roof should eventually become prismatically affected.

The calculations were carried out by means of logarithms of 8 decimals tables of 7 decimals being used only in calculating the mean hour at Greenwich at the moment of emersion of B A C 81 from the lunar disk, as observed at midnight between the 11th and 12th July, 1906, at Bujongolo, the last astronomic station in the district nearest to the Ruwenzori uplands.

The astronomic refraction r, corresponding to the considerable altitudes at which the astronomic observations were made during the journey, was calculated with Bessel's well-known formula:—

$$r = \log (a \tan z) + A (\log B + \log T) + \log \gamma,$$

neglecting the factor A, for apparent zenith distances z, under $77°$, and the factor λ, besides A, for apparent zenith distances less than $45°$. The values of the elements contained in the foregoing formula were deduced from Albrecht's tables, 1894 edition. But the Table 34f, which gives the value of log B, only comprises barometric pressures between 600 and 780 mm. (24 and 31 inches),

Appendix B.

whereas the expedition reached altitudes at which considerably lower pressures had to be recorded; hence, besides Albrecht's 34*f* table the following was also calculated, and is here inserted, as it may be found useful in other cases.

Barom.	log B.	Barom.	log B.	Barom.	log B.	Barom.	log B.	Barom.	log B.
mm.		mm.		mm.		mm.		mm.	
400 ·0	27387	440 ·0	23248	480 ·0	19469	520 ·0	15993	560 ·0	12774
1 ·0	27279	41 ·0	23149	81 ·0	19378	21 ·0	15909	61 ·0	12697
2 ·0	27170	42 ·0	23051	82 ·0	19288	22 ·0	15826	62 ·0	12619
3 ·0	27062	43 ·0	22953	83 ·0	19198	23 ·0	15743	63 ·0	12542
4 ·0	26933	44 ·0	22865	84 ·0	19108	24 ·0	15660	64 ·0	12465
5 ·0	26847	45 ·0	22752	85 ·0	19019	25 ·0	15577	65 ·0	12388
6 ·0	26740	46 ·0	22660	86 ·0	18929	26 ·0	15494	66 ·0	12311
7 ·0	26634	47 ·0	22562	87 ·0	18840	27 ·0	15412	67 ·0	12235
8 ·0	26527	48 ·0	22465	88 ·0	18751	28 ·0	15330	68 ·0	12158
9 ·0	26421	49 ·0	22368	89 ·0	18662	29 ·0	15247	69 ·0	12082
410 ·0	26315	450 ·0	22272	490 ·0	18573	530 ·0	15163	570 ·0	12006
11 ·0	26209	51 ·0	22175	91 ·0	18485	31 ·0	15084	71 ·0	11929
12 ·0	26103	52 ·0	22079	92 ·0	18396	32 ·0	15002	72 ·0	11853
13 ·0	25998	53 ·0	21983	93 ·0	18308	33 ·0	14920	73 ·0	11778
14 ·0	25893	54 ·0	21887	94 ·0	18220	34 ·0	14844	74 ·0	11702
15 ·0	25788	55 ·0	21792	95 ·0	18132	35 ·0	14758	75 ·0	11626
16 ·0	25684	56 ·0	21697	96 ·0	18045	36 ·0	14677	76 ·0	11551
17 ·0	25579	57 ·0	21601	97 ·0	17957	37 ·0	14596	77 ·0	11475
18 ·0	25475	58 ·0	21506	98 ·0	17870	38 ·0	14515	78 ·0	11406
19 ·0	25372	59 ·0	21412	99 ·0	17783	39 ·0	14434	79 ·0	11325
420 ·0	25268	460 ·0	21317	500 ·0	17696	540 ·0	14354	580 ·0	11250
21 ·0	25165	61 ·0	21223	1 ·0	17609	41 ·0	14273	81 ·0	11175
22 ·0	25062	62 ·0	21129	2 ·0	17523	42 ·0	14193	82 ·0	11101
23 ·0	24959	63 ·0	21035	3 ·0	17436	43 ·0	14113	83 ·0	11026
24 ·0	24856	64 ·0	20941	4 ·0	17350	44 ·0	14033	84 ·0	10952
25 ·0	24754	65 ·0	20848	5 ·0	17264	45 ·0	13933	85 ·0	10877
26 ·0	24652	66 ·0	20754	6 ·0	17178	46 ·0	13874	86 ·0	10803
27 ·0	24550	67 ·0	20661	7 ·G	17092	47 ·0	13794	87 ·0	10729
28 ·0	24449	68 ·0	20568	8 ·0	17007	48 ·0	13715	88 ·0	10655
29 ·0	24347	69 ·0	20476	9 ·0	16921	49 ·0	13636	89 ·0	10581
430 ·0	24246	470 ·0	20383	510 ·0	16836	550 ·0	13557	590 ·0	10508
31 ·0	24145	71 ·0	20291	11 ·0	16751	51 ·0	13478	91 ·0	10434
32 ·0	24045	72 ·0	20199	12 ·0	16666	52 ·0	13399	92 ·0	10361
33 ·0	23944	73 ·0	20107	13 ·0	16581	53 ·0	13320	93 ·0	10288
34 ·0	23844	74 ·0	20015	14 ·0	16497	54 ·0	13242	94 ·0	10214
35 ·0	23744	75 ·0	19924	15 ·0	16412	55 ·0	13164	95 ·0	10141
36 ·0	23644	76 ·0	19832	16 ·0	16328	56 ·0	13086	96 ·0	10068
37 ·0	23545	77 ·0	19741	17 ·0	16244	57 ·0	13088	97 ·0	9996
38 ·0	23444	78 ·0	19650	18 ·0	16160	58 ·0	12930	98 ·0	9923
39 ·0	23347	79 ·0	19559	19 ·0	16076	59 ·0	12852	99 ·0	9850

The value of log B, given in this table, is calculated with the formula:

$$\log B = \log (\{7 \cdot 12407 - 10\} b),$$

in which b is the barometric pressure in millimetres. (1 mm. = about $\frac{1}{25}$th inch.)

I.—Astronomic Observations.

The expedition of H.R.H. was supplied with four pocket chronometers at mean time, and before starting on the journey these were kept under control at the Hydrographic Institute. During this period of control, the absolute and daily corrections recorded for the said chronometers at 0ʰ of mean Greenwich time yielded the following results :—

Place.	Date, 1906.	Mean time.	Lange 56,509, K₁.	K₁.	Lange 56,520, K₂.	K₂.	Longines 560,229, K₃.	K₃.	Longines 560,234, K₄.	K₄.	Note.
Genoa	20 Feb.	11°·3	+9ˢ·87		−23ˢ·73		−12ˢ·43		−30ˢ·83		
				−0·27		−0·57		+1·77		+0·91	
,,	26 ,,	11·4	8·26		27·14		−1·79		25·39		
				−0·32		−0·78		1·65		1·91	
,,	3 Mar.	12·4	6·66		31·04		+6·46		15·84		
				−0·15		−0·69		2·08		1·12	
,,	8 ,,	14·1	5·93		34·47		+16·88		10·22		
				−0·41		+0·17		1·16		1·09	
,,	13 ,,	14·6	3·88		33·62		+22·68		4·77		
				−0·45		−0·60		2·88		0·50	
,,	19 ,,	14·0	1·20		37·20		39·95		1·80		
				+0·20		−1·07		2·67		0·32	
,,	24 ,,	12·4	2·22		42·53		53·32		0·18		
				+0·01		−0·96		2·13		0·40	
,,	29 ,,	11·0	2·29		47·31		1ᵐ 3ˢ·99		1·84		
				−0·11		+0·11		2·19		−0·13	
,,	3 Apr.	—	1·74		46·76		1 ·14·94		1·19		In pocket.
				+0·63		−0·29		2·81		+1·00	
,,	7 ,,	—	4·29		47·99		1 ·26·19		5·19		,, ,,
				+0·29		−0·07		5·34		−4·38	
Naples	14 ,,	—	6·38		47·52		2 ·04·68		36·88		In train.

After leaving Genoa, and more particularly during the voyage by steamer from Naples to Port Said, the chronometers were left unregulated. But at Port Said they were again set going, and on 20th April, 1906, compared with the chronometer at the Police Station, in order to record their absolute correction.

On the 26th of the same month another comparison was effected at Jibuti with the chronometer of *The Elphinstone* of the Indian Navy, and on 4th May, yet another with the chronometer of the Post Office at Mombasa. At Entebbe, thanks to steps previously taken, it was found possible on 12th May to make a fresh comparison by wire with Mombasa, so that by means of these two comparisons was obtained a first diurnal correction of the chronometers to be used in the subsequent calculations.

The elements of comparison appear in the record of the pocket chronometers included in the present Report. From it we find that at Mombasa, on 4th July, at noon, local time, there were the following absolute corrections on Greenwich mean time :

$$K_1 = + 3^h\ 16^m\ 34^s\cdot 9$$
$$K_2 = + 3\quad 26\quad\ 9\cdot 0$$
$$K_3 = + 3\quad\ 2\quad 10\cdot 5$$

Appendix B.

On 12th May, at Entebbe (noon at Mombasa), we obtained, by the above-mentioned telegraphic comparison, as correction on the mean Greenwich time :

$$K_1 = + \; 3^h \; 16^m \; 54^s \cdot 0$$
$$K_2 = + \; 3 \quad 25 \quad 52 \cdot 0$$
$$K_3 = + \; 3 \quad 2 \quad 10 \cdot 5$$

From these elements we get the following diurnal corrections for the three chronometers :

$$B_1 = - \; 2^s \cdot 762$$
$$B_2 = + \; 2 \cdot 215$$
$$B_3 = + \; 3 \cdot 437$$

The elements respecting the chronometer No. 4 have been omitted because on 7th May, when the party reached Entebbe, that chronometer was stolen.

The start for Ruwenzori was made at Entebbe, capital of the Uganda Protectorate, and on the march some astronomical observations were made in order to fix the position of some points which were generally those of encampments. Obviously it was not a case in which too much reliance could be placed on the Greenwich time, as indicated by the chronometers during the period of one month of rough travelling, that being about the time occupied in reaching Bujongolo, last point where were obtained astronomical observations, and where were begun the topographic operations for the survey of the Ruwenzori highlands. In order, however, to secure the greater or less efficiency of the chronometric observations, it was found expedient to observe, with the determination of the local time at Bujongolo, the emersion of B A C 81 from the lunar disk, with a view to calculating the hour of Greenwich time at the moment when the phenomenon was observed.

During the march the chronometers were carried on his person by H.R.H., who kept them carefully wrapped up. This expedient should have reduced to a minimum the influence of the changes of temperature, if, during the hours of rest, when being replaced in their own boxes, they had not had to feel the effects of the temperature inside the tent. Such effects, though little different from those of the atmosphere, always differed greatly from those due to contact with the human body. Still, when we consider that the period of repose was daily repeated for about the same length of time, it may be inferred that the daily recurring correction cannot have been affected by serious error due to this cause.

In any case it is to be regretted that of the three chronometers one alone displayed a sufficiently regular action, maintaining a fairly slight daily correction. This may easily be seen from the record of the chronometers.

I.—Astronomic Observations.

Here Nos. 2 and 3 point to irregularities in their movement. For this reason, and also because the daily comparisons were occasionally omitted, it was considered desirable to make use only of the indications of the No. 1 chronometer, which was in fact the one generally employed for the observations.

With the view of fixing, if only approximately, the daily correction of the No. 1 chronometer for the period of the journey, the calculation of the astronomic elements of Bujongolo was taken in hand, and here were recorded determinations of time between 11th and 28th June, taking the latitude at $\varphi = 0° 20' 16''$ N., roughly obtained from the already determined elements. The corrections of the No. 1 chronometer on the mean local time were for Bujongolo:

11th June,	3^h 40^m *	... Obs. No. 32	...	C_{tm} = +	5^h	15^m	$39^s.$	1
	3 43	„ 33		„ = +			39 ·	1
	3 48	„ 34		„ = +			35 ·	5
	3 49	„ 35		„ = +			34 ·	6
26th „	19 19	„ 37		„ = + 5		15	33 ·	2
	19 23	„ 38		„ = +			31 ·	9
27th „	19 38	„ 39		„ = + 5		15	29 ·	9
	19 43	„ 40		„ = +			32 ·	0
28th „	20 23	„ 41		„ = + 5		15	26 ·	8
	20 27	„ 42		„ = +			29 ·	4

From the mean of the results for 11th and 28th June respectively, we get, omitting the intermediate observations:

11th June,	3^h 45^m	C_{tm} = + 5^h 15^m $37^s.1$
28th „	20 25	„ = + 5 15 28 ·1

Interval 17^{days} 16^h 40^m Difference $9^s.0$

Hence: $B = - 0^s.509$.

The change occurring in the mean daily correction of this chronometer is seen to be considerable, if the value just found be compared with that previously obtained at Entebbe ($- 2^s.762$). But now we merely require an approximate value of the longitude of Bujongolo for the calculation of the emersion of B A C 81, and this will give us the absolute longitude of that same point. Hence we take the by no means arbitrary course of adopting, as mean daily correction of the No. 1 chronometer during the journey, the mean of the two daily corrections obtained at Entebbe and at Bujongolo, that is to say:

$$B_1 = - \frac{2^s.762 + 0^s.509}{2} = 1^s.635$$

---

* The date is astronomic, and the hour is referred to the mean local time.

Appendix B.

Referring the observations of the 26th, 27th and 28th June to the date of those of the 11th, and using the daily correction − 0^s·509 we get, by applying the just found mean daily correction 1^s·635, the following values for the longitude of Bujongolo :

11th June.—	Sun at W.	$\lambda = 1^h 59^m 53^s$· 8	E.G.
11th ,,	,, ,,	,, =	53 · 5	,,
11th ,,	,, ,,	,, =	49 · 9	,,
11th ,,	,, ,,	,, =	49 · 0	,,
26th ,,	Sun at E.	,, =	55 · 5	,,
26th ,,	,, ,,	,, =	54 · 3	,,
27th ,,	,, ,,	,, =	52 · 8	,,
27th ,,	,, ,,	,, =	54 · 9	,,
28th ,,	,, ,,	,, =	50 · 2	,,
28th ,,	,, ,,	,, =	52 · 8	,,

Grouping these results for each single day of observation we get :

Bujongolo.—11th June	... ...	$\lambda = 1^h 59^m 51^s$·5
26th ,,	... ...	,, =	54·9
27th ,,	... ...	,, =	53·8
28th ,,	... ...	,, =	51·5

Disregarding further considerations as to weight, and given the degree of approximation now required, the mean of these data is :

Bujongolo ... ... $\lambda = 1^h 59^m 52^s$·9 E.G.

This value is used in calculating the Greenwich time at the moment of the emersion of B A C 81 from the lunar disk, which phenomenon occurred on 11th July under most favourable conditions for observation. The determinations of the horary angle, obtained for this occasion with the view of ascertaining the state of the chronometer respecting the mean local time, gave the following results :

Bujongolo :

10th July,	21^h 18^m	....Obs. No. 47.	Sun at E.	$C_{tm} = + 5^h 15^m 33^s$· 0
10th ,,	21 20	,,	48	,, ,,	,, = +	32 · 3
11th ,,	20 01	,,	56	,, ,,	,, = + 5 15	34 · 9
11th ,,	20 03	,,	57	,, ,,	,, = +	33 · 7
11th ,,	20 07	,,	58	,, ,,	,, = +	32 · 4
11th ,,	20 11	,,	59	,, ,,	,, = +	32 · 9
12th ,,	18 08	,,	60	,, ,,	,, = + 5 15	34 · 4

312

I.— Astronomic Observations.

Although this completely agrees with the mean tenour of the other results, the last value is for the present neglected, and, after obtaining the mean of each day, we get as general mean:

11th July, 8^h 42^m ... ... $C_{tm} = +5^h$ 15^m $33^s\cdot1$.

From the observations taken at Bujongolo during the days following our arrival we had (*see* p. 311):

28th June, 20^h 25^m ... ... $C_{tm} = +5^h$ 15^m $28^s\cdot1$.

Hence for this interval of $12\cdot52$ days we obtain the diurnal correction:

$$K = +0^s\cdot398$$

with which we get:

12th July, 0^h of local time ... ... $C_{tm} = +5^h$ 15^m $33^s\cdot3$

Moment of Occultation ... ... ,, $= +5^h$ 15^m $33^s\cdot2$

With this element and with the approximate longitude already obtained, we proceed to a first calculation of mean Greenwich time at the moment of emersion of B A C 81 from the lunar disk, the moment when the No. 1 chronometer indicated 10^h 14^m 4^s (Obs. No. 55). From the first approximation we got:

Bujongolo ... ... $\lambda = 1^h$ 59^m $59^s\cdot2$ E.G.

The calculation for a second approximation, in which account was also taken of the terms of second order, only very slightly modified the result. Thus:

Bujongolo ... ... $\lambda = 1^h$ 59^m $59^s\cdot33$ E.G.

As, however, the value of the longitude thus obtained might be seriously affected by even a slight error in the lunar co-ordinates given by the ephemerides, we consulted some astronomic observers in order to ascertain whether, about the time when the expected occultation took place, any observations of lunar culminations had concurrently been made. This was done in order to introduce into the calculation the error of the position of the moon.

Prof. Millosevich, Director of the Observatory of the Collegio Romano, in Rome, having undertaken the determination of the longitude of Tripoli, where the astronomer, Dr. Bianchi, was observing transits of the moon at meridian, proceeded to take observations of lunar culminations at the Collegio Romano from the 2nd to the 7th July, 1906. From these he obtained for 11th July—time of the occultation—a correction for the right ascension of the moon $= +0^s\cdot18$, and this agrees perceptibly with that communicated to us by Greenwich for the same date $= +0^s\cdot20$.

It may be mentioned that Greenwich also supplied us with the correction for that date of the lunar declination $= +1^s\cdot8$.

Appendix B.

Hence the same Prof. Millosevich advised us to assume with full confidence the corrections for the lunar co-ordinates received from Greenwich, and these yielded the longitude for Bujongolo :

$$\lambda = 2^h \ 0^m \ 6^s{\cdot}3 \ \text{E.G.}$$

This again agrees closely with the value $2^h \ 0^m \ 6^s{\cdot}0$ East Greenwich, obtained by Prof. Millosevich, who was also good enough to make the same calculation.

The latitude was obtained from two meridian altitudes and from two series of circummeridians (Obs. Nos. 38, 43, 44, 46, and 49 to 54), observed partly by H.R.H. and partly by Commander Cagni.   Between the results of the two observers there occurs a considerable difference, the origin of which may be attributed to some anomaly of refraction.   In fact, H.R.H. was in this instance induced to depart from his practice of observing the lower limb of the sun, owing to an unusual optic phenomenon which caused him to notice on the lower edge of the reflected image a false limb which would not have allowed a good observation.   Hence the discrepancy in the results is to be attributed to this particular state of the atmosphere.   Therefore, in order to prevent the observations of H.R.H., which were the more numerous, from too greatly influencing the results, the mean of the circummeridian series was first obtained, and the resulting value taken as a mean with the results of the meridian observations.   The several values thus obtained are :

17th June. —Meridian   — Commander Cagni $\phi = 0°$ 19′ 50″ N.
 9th July. —   ,,   — H.R.H.   ,, $= 0$ 20   55   ,,
10th   ,,   —Circummerid.— Commander Cagni ,, $= 0$ 19   52   ,,
11th   ,,   —Meridian   — H.R.H.   ,, $= 0$ 20   54   ,,

These data yielded for

Bujongolo   ...   ... $\phi = 0°$ 20′ 23″ N.

With the longitude of Bujongolo is obtained the absolute correction of the chronometer for the period of arrival at that encampment, and subsequently the mean daily correction of the same chronometer for the period occupied by the journey.

Thus was obtained :

Bujongolo.—19th June, $3^h \ 45^m$ ... $C_{tm} = + 5^h \ 15^m \ 37^s{\cdot}1$
    ,,   ,,   ,,   ... $\lambda = + 2$ 0   6 ·3
                                         $K_1 = + 3$ 15   30 ·8

This absolute correction on mean Greenwich time corresponds with the date of 11th June at $3^h \ 45^m$ of mean local time.   And as at Entebbe on 11th May, at $23^h \ 31^m$ of mean local time, we had $K_1 = 3^h \ 16^m \ 34^s{\cdot}$ 9, we shall

I.— Astronomic Observations.

get, taking account of the difference of longitude between Bujongolo and Entebbe ($+$ 9^m 45^s), the daily mean correction $K_1 = -$ 2$^s \cdot$ 123, which we shall utilize for the determinations of position made in the period from 11th May to 11th June.

Owing to an unforeseen circumstance, on the return journey, and after the arrival of the expedition at Fort North Portal, the No. 1 chronometer underwent, like the others, a perceptible change in its movement. This was due to a considerable delay which occurred in winding it, so that once it was necessary to proceed to the revision of the longitude of Fort Portal which had been determined on the outward journey. On the return, the conveyance of Greenwich time from Bujongolo will be limited to this intermediate point, since, owing to the above-mentioned change in the movement of the chronometer, it would be impossible to convey said time to Entebbe for purposes of control.

Retaining the value of the daily correction just found ($K_1 = -$ 2$^s \cdot$ 123) as a quantity proportional to the time, and with the (approximate) latitude of Fort Portal $= 0°$ 39$'$ 20$''$ N., we get the value of the longitude from four series of altitudes (Obs. Nos. 17, 18, 19, and 20), obtaining :

$$\text{Fort Portal.—31st May.} \quad \lambda = 2^h \; 1^m \; 32^s \cdot 2 \text{ E.G.}$$

,,	,,	,, $=$	31 ·8 ,,
,,	,,	,, $=$	31 ·1 ,,
,,	,,	,, $=$	34 ·7 ,,

and as mean :

$$\text{Fort Portal} \quad \ldots \quad \ldots \quad \lambda = 2^h \; 1^m \; 32^s \cdot 5 \text{ E.G.}$$

a value which is adopted as the longitude of said place.

The latitude of Fort Portal is obtained from a series of three circum-meridians observed on 31st May, and from meridian altitudes of 22nd and 28th July (Obs. Nos. 21, 22, 23, 75, and 86). The mean of the five results yielded for :

$$\text{Fort Portal} \quad \ldots \quad \ldots \quad \phi = 0° \; 39' \; 28'' \text{ N.}$$

a value which differs little from that employed for the calculation of the longitude.

Using the longitude just found, we get the absolute correction of No. 1 chronometer at Fort Portal (return journey) by means of eight series of observations, as under :

Fort Portal.—21st July,	4^h 11^m	...Obs. No.	73;	$K_1 = +$ 3^h 15^m	16$^s \cdot$ 1
,,	4 18	,,	74;	,, $=$	15 · 9
,,	19 53	,,	76;	,, $=$	17 · 5
,,	19 55	,,	77;	,, $=$	16 · 5

315

Appendix B.

Fort Portal. — 22nd July, $19^h 53^m$ ... Obs. No. 78 ; $K_1 = 3^h 15^m 18^{s\cdot}8$

22nd ,, 19 55 ,, 79 ; ,, = 20·1

23rd ,, 20 6 ,, 80 ; ,, = 22·6

23rd ,, 20 8 ,, 81 ; ,, = 22·2

Of this the mean for the double series required by the above-mentioned inversion of the glass roof of the artificial horizon is :

21st July, $4^h 15^m$ ... $K_1 = +3^h 15^m 16^{s\cdot}0$

21st ,, 19 54 ,, = + 17·0

22nd ,, 19 54 ,, = + 19·5

23rd ,, 20 7 ,, = + 22·4

Referring these values to the date coinciding with the first of them, and noting the hour indicated by the chronometer, we get :

21st July (civil)—(p.m.) : $t_c = 10^h 57^m 49^s ... K_1 = 3^h 15^m 15^s\cdot8$

as the mean on Greenwich mean time.

From the determinations of the time at Bujongolo on the 10th, 11th, and 12th July (astronomic dates), the results of which have been given at p. 312, we obtain the absolute correction of No. 1 chronometer. Referring all the values to the date of the last, and passing from the absolute correction of the chronometer to the absolute correction on Greenwich, we get :

13th July (civil)—(a.m.) : $t_c = 2^h 51^m 6^s ... K_1 = 3^h 15^m 28^s\cdot4$

from which in the interval between the 13th and 21st July the daily correction of the chronometer is found to be :

$$K_1 = -1^s\cdot521.$$

As already stated at p. 315, on the return journey the chronometers at Fort Portal varied considerably through lack of control, so that, before leaving this place, we proceeded to the determination of their correction by means of six series of altitudes, the results being :

27th July.—Obs. No. 82 ; $t_c = 10^h 20^m 11^s$ ... $K_1 = +3^h 32^m 57^{s\cdot}6$

27th ,, ,, 83 ; ,, = 10 23 0 ,, = 51·0

28th ,, ,, 84 ; ,, = 2 21 0 ,, = 59·5

28th ,, ,, 85 ; ,, = 2 23 8 ,, = 59·9

31st ,, ,, 87 ; ,, = 10 25 44 ,, = 64·8

31st ,, ,, 88 ; ,, = 10 27 53 ,, = 64·7

The disagreement of the second series induced us to abandon it, the influence of some error in the observations being obvious.

I.—Astronomic Observations.

Referring the daily values of the absolute correction to the mean date of the values of the last double series we get :

31st July (civil) (p.m.), $t_c = 10^h 26^m 48^s ... K_1 = + 3^h 33^m 5^{s} \cdot 0$

which represents the absolute correction of the chronometer on Greenwich time before starting on the return journey from Fort Portal to Entebbe. On reaching the latter place it was found impossible to get a new telegraphic comparison with Mombasa, as on the outward journey. Hence proceeded to the determination of the absolute correction of the chronometer, using for Entebbe the longitude $2^h 9^m 47^s$ East Greenwich given us by the competent local authority. The result was :

16th August.—Obs. No. 110 ; $t_c = 10^h 29^m 7^s$ ... $K_1 = + 3^h 33^m 29^{s} \cdot 9$
16th   ,,   ,,   111 ; ,, $= 10 \quad 31 \quad 13$,, $=$ $28 \cdot 5$
17th   ,,   ,,   112 ; ,, $= 2 \quad 43 \quad 55$,, $=$ $25 \cdot 2$
17th   ,,   ,,   113 ; ,, $= 2 \quad 46 \quad 1$,, $=$ $23 \cdot 8$

We see *à priori* that in this interval of little over 16 hours the movement of the chronometer indicates a strong variation, such as had never occurred during the whole journey. Instead of taking the mean of these values, it was thought expedient to use the results alone of the two series of 16th August observed immediately after the arrival at Entebbe. From these we get :

16th August (civil), (p.m.) : $10^h 30^m 10^s ... K_1 = + 3^h 33^m 29^{s} \cdot 2$

so that the daily correction of the chronometer to be used in the interval from 31st July to 15th August was :

$$K_1 = + 1^{s} \cdot 510.$$

The question now was to see what degree of confidence might be placed in the daily corrections which had so far been obtained. From the fact that the longitude of a few points was determined both on the outward and the return journey, we were offered a means of control which, if it stood alone, would not be absolutely safe, since it was always possible that the errors by which the accepted daily corrections might be affected might be such, in magnitude and sign (plus or minus), as to lead to longitudinal results apparently concordant though really very incorrect. As, however, there were several points determined under such conditions, so that in some cases we could ascertain the degree of concordance in the longitudinal results, from this might be inferred both the practical value of the daily corrections that had been adopted, and the measure of confidence that might be placed in the positions obtained from the astronomic observations.

Appendix B.

The position of Ibanda, a place lying between Bujongolo and Fort Portal, was determined both on going and returning. From four meridian altitudes (Obs. Nos. 27, 65, 66, and 67), we obtained for this point the latitude $\phi = 0° 19' 59''$ N., and from this were deduced the following longitudinal values :—

On the outward journey :

$$\text{Ibanda Obs. No. 28......... } \lambda = 2^h 0^m 44^s \cdot 0 \text{ E.G.}$$
$$\text{,, \quad ,, \quad } 29......... \text{ ,, } = \quad 43 \cdot 1 \text{ ,,}$$
$$\text{Mean ,, } = 2^h 0^m 43 \cdot 5 \text{ ,,}$$

On the return journey :

$$\text{Ibanda.—Obs. No. 62 } ... \quad \lambda = 2^h 0^m 43^s \cdot 9 \text{ E.G.}$$
$$\text{,, \quad } 63 \quad ... \quad \text{,, } = \quad 41 \cdot 5 \text{ ,,}$$
$$\text{,, \quad } 64 \quad ... \quad \text{,, } = \quad 42 \cdot 5 \text{ ,,}$$
$$\text{,, \quad } 68 \quad ... \quad \text{,, } = \quad 42 \cdot 7 \text{ ,,}$$
$$\text{,, \quad } 69 \quad ... \quad \text{,, } = \quad 43 \cdot 5 \text{ ,,}$$
$$\text{Mean ,, } = 2^h 0^m 42^s \cdot 8 \text{ ,,}$$

Such is the agreement between these two results that we may even disregard all considerations as to the weights to be adopted for the values obtained, whether as regards the number of concurrent observations, or the length of time during which Greenwich time had to be conveyed. In this case there intervened 22 days for the determination of longitude on going (that is, Greenwich time was conveyed for 22 days), compared with a mean of about 5 days of conveyance for the determination made on our return.

Moreover, given the degree of accuracy that may be required, allowing for the available means and the limited time at the disposal of the expedition, we found it advisable to adopt as the value of the longitude of Ibanda the mean of the two results, namely :

$$\text{Ibanda......... } \lambda = 2^h 0^m 43^s \cdot 2 \text{ E.G.}$$

In the district between Entebbe and Port Portal there are two other points which offered the same conditions, and which consequently contributed to supply means of control. For Kichiomi, which is one of these two points, we obtained by the observation of a meridian altitude both going and returning (Obs. Nos. 10 and 99) the following result :

$$\text{Kichiomi......... } \phi = 0° 31' 20'' \text{ N.}$$

Adopting this value for the calculation of longitude we obtained on going :

$$\text{Kichiomi.—Obs. No. 11......... } \lambda = 2^h 4^m 27^s \cdot 3 \text{ E.G.}$$

I.—Astronomic Observations.

On the return :

Kichiomi.—Obs. No. 100........$\lambda = 2^h\ 4^m\ 26^{s.}\ 0$ E.G.

$\qquad$,,　,, 101　　　,,　　　 $25 \cdot 9$　,,

$\qquad$ Mean........$\lambda = 2^h\ 4^m\ 26^{s.}\ 0$　,,

Here also the agreement between the two results is satisfactory, and for the reasons already stated we retain as definite value the mean of the two results, as under :

$\qquad$ Kichiomi......$\lambda = 2^h\ 4^m\ 26^{s.}\ 7$.

An analogous process is taken for Muyongo, where the latitude $\varphi = 0°\ 30'\ 41''$ N. was obtained by two circummeridians (Obs. Nos. 12 and 13) observed on going ; introducing this value in the calculation of the longitude, for which there are two series of altitudes on going and two on returning, we get :

Going :

$\qquad$ Misongo.—Obs. No. 14........$\lambda = 2^h\ 3^m\ 56^{s.}\ 5$ E.G.

$\qquad$,,　　 15　　　,, $=$　　 $55^{s.}\ 8$　,,

Returning :

$\qquad$ Misongo.—Obs. No. 97........$\lambda = 2^h\ 3^m\ 55^{s.}\ 4$ E.G.

$\qquad$,,　　 98　　　,, $=$　　 $54^{s.}\ 6$　,,

or taking the simple mean :

$\qquad$ Misongo...........$\lambda = 2^h\ 3^m\ 55^{s.}\ 6$ E.G.

Thus we get a third test regarding the practical value of the daily corrections adopted for the chronometer.

And since the results of longitude were repeatedly concordant in a measure greater than had been expected, we may proceed to the calculation of the elements of position for all the other points determined during the expedition, being confident of incurring no serious errors.

Bujongo (near Lake Isolt).—The latitude was obtained by a meridian altitude observed on the outward journey (Obs. No. 1), and the longitude by two series of altitudes also on going (Obs. Nos. 2 and 3) :

$\qquad$ 19th May $\varphi = 0°\ 25'\ 44''$ N............$\lambda = 2^h\ 4^m\ 53^{s.}\ 4$ E.G.

$\qquad$,,　　　　　　　　　　　　 ,, $= 2\ \ 7\ \ 54 \cdot 5$　,,

$\qquad$ Bujongo　$\varphi = 0°\ 25'\ 44''$ N............$\lambda = 2^h\ 7^m\ 53^{s.}\ 9$　,,

$\qquad$ (Lake Isolt)　　　　　　　　 ,, $= 31°\ 58'\ 28''$　,,

Appendix B.

Bimbye.—Latitude determined by two meridian altitudes (going and returning, Obs. Nos. 4 and 107); longitude by two series of altitudes (return, Obs. Nos. 108 and 109).

20th May	$\phi = 0°\ 31'\ 56''$ N.	
10th August	„ $= 0\ \ 31\ \ 57$ „	
11th „		$\lambda = 2^h\ 7^m\ 26^{s\cdot}1$ E.G.
11th „		„ $= 2\ \ 7\ \ \ 26\cdot4$ „
Bimbye	$\phi = 0°\ 31'\ 56''$ N. ...	$\lambda = 2^h\ 7^m\ 26^{s\cdot}3$ „
		„ $= 31°\ 51'\ 34''$ „

Kijemula.—Latitude determined by a meridian altitude (return, Obs. No. 106), the longitude being estimated at $2^h\ 6^m$ E.G.

Kijemula.—9th August $\phi = 0°\ 35'\ 55''$ N.

Muduma.—Latitude obtained by a meridian altitude (going, Obs. No. 5), and longitude by two series of altitudes (going, Obs. Nos. 6 and 7):

23rd May	$\phi = 0°\ 36'\ 19''$ N. ...	$\lambda = 2^h\ 5^m\ 40^{s\cdot}3$ E.G.
		„ $= 2\ \ 5\ \ \ 40\cdot9$ „
Muduma	$\phi = 0°\ 36'\ 19''$ N. ...	$\lambda = 2\ \ 5\ \ \ 40\cdot6$ „
		„ $= 31°\ 25'\ 9''$ „

Kasiba.—Position determined on the return: Latitude, by a meridian altitude (Obs. No. 103); longitude by two series of altitudes (Obs. Nos. 104 and 105):

8th August	$\phi = 0°\ 40'\ 34''$ N. ...	$\lambda = 2^h\ 5^m\ 53^{s\cdot}2$ E.G.
8th „		„ $= 2\ \ 5\ \ \ 50\cdot8$ „
Kasiba	$\phi = 0°\ 40'\ 34''$ N. ...	$\lambda = 2^h\ 5^m\ 52^{s\cdot}0$ „
		„ $= 31°\ 28'\ 0''$ „

Lwamutukuza.—Latitude obtained by two meridian altitudes, one going the other returning (Obs. Nos. 8 and 102); longitude by a series of altitudes going (Obs. No. 9):

24th May	$\phi = 0°\ 31'\ 4''$ N. ...	$\lambda = 2^h\ 5^m\ 16^{s\cdot}5$ E.G.
7th August	„ $= 0\ \ 30\ \ 27$ „	
Lwamutukuza	$\phi = 0°\ 30'\ 45''$ N. ...	$\lambda = 2^h\ 5^m\ 16^{s\cdot}5$ „
		„ $= 31°\ 19'\ 7''$ „

I.—Astronomic Observations.

Kaibo.—Latitude determined by a circummeridian altitude going and four returning (Obs. Nos. 18, 91, 92, 93, and 94); longitude results from two series of altitudes on return (Obs. Nos. 95 and 96):

27th May	$\phi = 0°\ 29'\ 56''$ N.		
2nd August	„ $= 0\ \ 30\ \ 36$ „	...	$\lambda = 2^h\ 3^m\ 7^{s\cdot}8$ E.G.
2nd „	„ $= 0\ \ 30\ \ 09$ „	...	„ $= 2\ \ 3\ \ 7 \cdot 9$ „
2nd „	„ $= 0\ \ 29\ \ 26$ „		
2nd „	„ $= 0\ \ 30\ \ 14$ „		
Kaibo	$\phi = 0°\ 30'\ 4''$ N.	...	$\lambda = 2^h\ 3^m\ 7^{s\cdot}9$ E.G.
			„ $= 30°\ 46'\ 58''$ „

Butiti.—Using the latitude estimated at $\phi = 0°\ 39'\ 30''$, the longitude is obtained by two series of altitudes observed on the return (Obs. Nos. 89 and 90):

1st August	...	...	...	$\lambda = 2^h\ 2^m\ 34^{s\cdot}3$ E.G.
1st „	...	...	...	„ $= 2\ \ 2\ \ 34 \cdot 8$ „
Butiti ...	...	...	...	$\lambda = 2^h\ 2^m\ 37^{s\cdot}5$ „
				„ $= 30°\ 38'\ 37''$ „

Duwona.—With the approximate value $\lambda = 2^h\ 1^m\ 17^s$ East Greenwich, the latitude is calculated by a meridian altitude observed on going (Obs. No. 24):

1st June.—Duwona......... $\phi = 0°\ 33'\ 25''$ N.

Butanuka.—Latitude determined on the return by a meridian altitude (Obs. No. 70); and longitude by two series of altitudes (Obs. Nos. 71 and 72):

20th July	$\phi = 0°\ 26'\ 33''$ N.	...	$\lambda = 2^h\ 1^m\ 4^{s\cdot}1$ E.G.
20th „			„ $= 2\ \ 1\ \ 4 \cdot 6$ „
Butanuka	$\phi = 0°\ 26'\ 33''$ N.	...	$\lambda = 2^h\ 1^m\ 4^{s\cdot}4$ „
			„ $= 30°\ 16'\ 6''$ „

Bihunga.—Longitude obtained by two series of altitudes on going, using $\phi = 0°\ 20'\ 20''$ N. obtained from the maps (Obs. Nos. 30 and 31):

4th June... ... ... $\lambda = 2^h\ 0^m\ 27^{s}\cdot 0$ E.G.

„ $= 2\ \ 0\ \ 27 \cdot 2$ „

Nakitawa.—Obtaining from the maps the latitude $\phi = 0°\ 20'\ 20''$ N., the longitude was had by a single series of altitudes at the artificial horizon without roof (Obs. No. 61):

Nakitawa $\lambda = 2^h\ 0^m\ 19^{s}\cdot 6$ E.G.

„ $= 30°\ 4'\ 54''$ „

Appendix B.

Summary of the geographical positions obtained by astronomic observations taken during the outward and return journey between Entebbe and Bujongolo.

Places.	Geographical Position.					
	Latitude North.			Longitude E.G.		
	°	′	″	°	′	″
Bujongo (Lake Isolt)	0	25	44	31	58	28
Bimbye	0	31	56	31	51	34
Kijemula	0	35	55			
Muduma	0	36	19	31	25	9
Kasiba	0	40	34	31	28	0
Lwamutukuza	0	30	45	31	19	7
Kichiomi	0	31	20	31	6	40
Muyongo	0	30	41	30	58	54
Kaibo	0	30	4	30	46	58
Butiti				30	38	37
Fort Portal	0	39	28	30	23	7
Duwona	0	33	25			
Butanuka	0	36	33	30	16	6
Kasongo				30	15	9
Ibanda	0	19	59	30	10	48
Bihunga				30	6	46
Nakitawa				30	4	54
Bujongolo	0	20	23	30	1	34

REGISTER OF THE
ASTRONOMIC OBSERVATIONS.

Appendix B.

Civil Date, 1906.	Numerical Order.	Observer.	Place.	Latitude N.	Longitude E.G.	Barom. mm.	Thermom. annexed (Celsius).	Temp. of the Air (Celsius).	Position Lid of the Artificial Horizon.
				° ′ ″	h. m. s.		°	°	
19th May	1	H.R.H.	Bujongo (nr.Lake Isolt)	....	2 7 54	663 ·9	26 ·0	31 ·0	0A
,,	2	,,	,,	0 25 44	....	661 ·0	24 ·0	29 ·0	0A △
,,	3	,,	,,	,,	....	,,	,,	,,	0A ▽
20th May	4	,,	Bimbye	....	2 7 26	666 ·8	26 ·0	28 ·0	0A
23rd May	5	,,	Muduma	....	2 5 41	658 ·0	22 ·0	23 ·5	0A
,,	6	,,	,,	0 36 19	....	654 ·5	23 ·0	24 ·5	0A △
,,	7	,,	,,	,,	....	,,	,,	,,	0A ▽
24th May	8	,,	Lwamutu-kuza	....	2 5 16	657 ·3	22 ·0	29 ·0	0A
,,	9	,,	,,	0 30 45	....	656 ·0	22 ·0	22 ·0	0A △
25th May	10	,,	Kichiomi	. ...	2 4 27	653 ·55	28 ·0	24 ·5	0A
,,	11	,,	,,	0 31 20	....	652 ·0	24 ·0	24 ·0	0A △
26th May	12	,,	Muyongo	....	2 3 56	658 ·0	25 ·0	27 ·0	0A
,,	13	,,	,,	....	,,	,,	,,	,,	....
,,	14	,,	Wonyongo	0 30 41	....	657 ·0	25 ·0	25 ·0	0A △
,,	15	,,	,,	,,	....	,,	,,	,,	0A ▽

I.—Astronomic Observations.

ASTRONOMIC OBSERVATIONS.

Conditions under which the Aster was observed.	Instrumental Height.	Instrument of Correction.	Hour of the Chronometer.	Number of the Chronometer.	Absolute Correction of the Chronometer.	Results.
	° ′ ″	′ ″	h. m. s.		h. m. s.	
Sun ☉ PS	141 6 20	+1 0	....	....	....	φ=0° 25′ 44″ N.
Sun ☉ to W	53 0 0 52 40 0 52 20 0	+1 20	10 38 31·0	1	+3 16 19·7	h. m. s. λ=2 7 53·4 E.G.
Sun ☉ to W	52 0 0 51 40 0 51 20 0	+1 20	10 40 36·0 41 22·5 42 7·0	1	+3 16 19·7	h. m. s. λ=2 7 54·5 E.G.
Sun ☉ PS	140 52 30	+1 20	....	....	....	φ=0° 31′ 56″ N.
Sun ☉ PS	139 47 20	+1 40	....	....	....	φ=0° 36′ 19″ N.
Sun ☉ to W	50 20 0 50 0 0 49 40 0	+1 45	10 46 31·0 47 17·0 47 58·5	1	+3 16 11·2	h. m. s. λ=2 5 40·3 E.G.
Sun ☉ to W	49 20 0 49 0 0 48 40 0	+1 45	10 48 42·0 49 24·5 50 8·0	1	+3 16 11·2	h. m. s. λ=2 5 40·9 E.G.
Sun ☉ PS	139 14 20	+1 0	....	....	....	φ=0° 31′ 4″ N.
Sun ☉ to W	43 20 0 43 0 0 42 40 0	+1 0	11 1 57·0 2 48·0 3 18·0	1	+3 16 9·0	h. m. s. λ=2 5 16·5 E.G.
Sun ☉ PS	138 50 20	+2 0	....	....	....	φ=0° 30′ 47″ N.
Sun ☉ to W	52 0 0 51 40 0	+1 0	10 43 59·0 44 43·0	1	+3 16 6·9	h. m. s. λ=2 4 27·3 E.G.
Sun ☉ Circumm.	138 28 40	+1 50	16 35 25·0	1	+3 16 5·2	φ=0° 30′ 51″ N.
„	138 25 40	„	16 40 30·0	1	+3 16 5·2	φ=0° 30′ 30″ N.
Sun ☉ to W	44 0 0 43 40 0 43 20 0	+2 10	11 1 48·5 2 33·0 3 18·0	1	+3 16 4·8	h. m. s. λ=2 3 36·5 E.G.
Sun ☉ to W	43 0 0 42 40 0 42 0 0	+2 10	11 4 00·0 4 43·0 5 29·0	1	+3 16 4·8	h. m. s. λ=2 3 55·8 E.G.

Appendix B.

Civil Date, 1906.	Numerical Order.	Observer.	Place.	Latitude N.	Longitude E.G.	Barom. mm.	Thermom. annexed (Celsius).	Temp. of the Air (Celsius).	Position Lid of the Artificial Horizon.
				° ′ ″	h. m. s.		°	°	
27th May	16	H.R.H.	Kaibo	....	2 3 8	652·0	24·0	23·6	0A
31st May	17	,,	Fort Portal	0 39 28	....	640·0	22·0	21·0	0A △
,,	18	,,	,,	,,	....	,,	,,	,,	0A ▽
,,	19	,,	,,	,,	....	,,	,,	,,	0A △
,,	20	,,	,,	,,	....	,,	,,	,,	0A ▽
,,	21	,,	,,	....	2 1 32	641·0	22·0	21·0	0A
,,	22	,,	,,	....	,,	,,	,,	,,	,,
,,	23	,,	,,	....	,,	,,	,,	,,	,,
1st June	24	,,	Duwona	....	2 1 17	636·4	24·0	24·0	0A
2nd June	25	,,	Kazongo	0 21 30	....	649·0	26·0	27·0	0A △
,,	26	,,	,,	,,	....	,,	,,	,,	0A ▽
3rd June	27	,,	Ibanda	....	2 0 43	652·8	26·5	25·5	0A
,,	28	,,	,,	0 19 59	....	650·0	25·0	23·0	0A △
,,	29	,,	,,	,·	....	,,	,,	,,	0A ▽

I.—Astronomic Observations.

Conditions under which the Aster was observed.	Instrumental Height.	Instrument of Correction.	Hour of the Chronometer.	Number of the Chronometer.	Absolute Correction of the Chronometer.	Results.
	° ′ ″	′ ″	h. m. s.		h. m. s.	
Sun ☉ Circumm.	138 6 0	+ 1 40	6 36 29 ·0	1	+ 3 16 8 ·2	φ = 0° 29′ 56″ N.
Sun ☉ to E	36 14 40 36 35 40 36 52 20 37 2 40	+ 2 20	1 58 5 ·0 58 50 ·0 59 28 ·0 59 50 ·0	1	+ 3 15 54 ·9	h. m. s. λ = 2 1 32 ·2 E.G.
Sun ☉ to E	37 20 40 37 45 40 38 4 20 38 15 40	+ 2 20	2 0 28 ·0 1 24 ·0 2 3 ·0 2 29 ·5	1	+ 3 15 54 ·9	h. m. s. λ = 2 1 31 ·8 E.G.
Sun ☉ to E	54 20 0 54 40 0 55 0 0 55 20 0	+ 2 20	2 37 37 ·0 38 19 ·5 39 6 ·0 39 49 ·5	1	+ 3 15 54 ·9	h. m. s. λ = 2 1 31 ·1 E.G.
Sun ☉ to E	60 9 0 60 18 20 61 14 20	+ 2 20	2 50 22 ·0 50 44 ·0 52 47 ·0	1	+ 3 15 54 ·8	h. m. s. λ = 2 1 34 ·7 E.G.
Sun ☉ Circumm.	134 31 40	+ 1 0	6 8 50 ·0	1	+ 3 15 55 ·0	φ = 0° 38′ 45″ N.
,,	134 54 20	+ 1 0	6 11 0 ·0	1	,,	φ = 0° 39′ 49″ N.
,,	135 1 40	+ 1 0	6 11 59 ·0	1	,,	φ = 0° 39′ 7″ N.
Sun ☉ PS	136 38 40	+ 2 20	....	....	....	φ = 0° 33′ 25″ N.
Sun ☉ to W	42 20 0 42 3 40 41 51 20	+ 1 20	11 8 35 ·0 9 12 ·0 9 38 ·0	1	+ 3 15 49 ·9	h. m. s. λ = 2 1 0 ·8 E.G.
Sun ☉ to W	41 28 20 41 18 0 41 51 40	+ 1 20	11 10 29 ·0 10 52 ·0 11 51 ·0	1	+ 3 15 49 ·9	h. m. s. λ = 2 1 0 ·4 E.G.
Sun ☉ PS	135 39 20	+ 1 30	....	....	....	φ = 0° 19′ 26″ N.
Sun ☉ to W	44 0 0 43 40 0 43 20 0	+ 1 30	11 5 16 ·0 6 1 ·0 6 44 ·0	1	+ 3 15 47 ·8	h. m. s. λ = 2 0 44 ·0 E.G.
Sun ☉ to W	43 0 0 42 40 0 42 20 0	+ 1 30	11 7 30 ·0 8 12 ·0 8 56 ·0	1	+ 3 15 47 ·7	h. m. s. λ = 2 0 43 ·1 E.G.

Appendix B.

Civil Date, 1906.	Numerical Order.	Observer.	Place.	Latitude N.	Longitude E.G.	Barom. mm.	Thermom. annexed (Celsius).	Temp. of the Air (Celsius).	Position Lid of the Artificial Horizon.
				° ′ ″	h. m. s.		°	°	
4th June	30	H.R.H.	Bihunga	0 20 20	....	611·0	20·0	20·0	0A △
,,	31	,,	,,	,,	....	,,	,,	,,	0A ▽
11th June	32	,,	Bujongolo	0 20 23	....	488·4	6·0	4·0	0A △
,,	33	,,	,,	,,	....	,,	,,	,,	0A ▽
,,	34	,,	,,	,,	....	,,	,,	,,	0A △
,,	35	,,	,,	,,	....	,,	,,	,,	0A ▽
17th June	36	Com. Cagni	,,	....	2 0 6	488·3	5·0	5·0	0A
27th June	37	,,	,,	0 20 23	....	489·2	5·0	6·0	0A △
,,	38	,,	,,	,,	....	,,	,,	,,	0A ▽
28th June	39	,,	,,	,,	....	490·0	7·0	6·8	0A △
,,	40	,,	,,	,,	....	,,	,,	,,	0A ▽

I.—Astronomic Observations.

Conditions under which the Aster was observed.	Instrumental Height.	Instrument of Correction.	Hour of the Chronometer.	Number of the Chronometer.	Absolute Correction of the Chronometer.	Results.
	° ′ ″	′ ″	h. m. s.		h. m. s.	
Sun ☉ to W	47 20 0	+0 50	10 58 22 ·0	1	+3 15 45·7	h. m. s.
	47 0 0		59 6 ·0			} λ=2 0 27 ·0 E.G.
	46 40 0		59 49 ·0			
Sun ☉ to W	46 20 0	+0 50	11 0 34 ·0	1	+3 15 45 ·7	h. m. s.
	46 0 0		1 18 ·0			} λ=2 0 27 ·2 E.G.
	45 40 0		2 2 ·0			
Sun ☉ to W	63 33 20	+1 25	10 23 26 ·0	1	....	h. m. s.
	63 14 40		24 9 ·0			} C_{tm}= +5 15 39 ·1
	62 50 20		25 4 ·0			
Sun ☉ to W	62 19 40	+1 25	10 26 11 ·0	1	....	h. m. s.
	62 8 40		26 39 ·0			} C_{tm}= +5 15 39 ·1
	61 54 40		27 7 ·5			
Sun ☉ to W	60 0 0	+1 25	10 31 30 ·0	1	....	h. m. s.
	59 40 0		32 14 ·0			} C_{tm}= +5 15 35 ·5
Sun ☉ to W	59 20 0	+1 25	10 33 0 ·0	1	....	h. m. s.
	59 0 0		33 45 ·0			} C_{tm}= +5 15 34 ·6
Sun ☉ PS	133 22 23	+2 20	....	....	....	φ=0° 19′ 50″ N.
Sun ☉ to E	34 21 0	+3 40	2 2 36 ·5	1	....	
	34 38 40		3 17 ·5			
	34 59 0		4 0 ·0			} h. m. s.
	35 14 30		4 36 ·0			C_{tm}= +5 15 33 ·2
	35 38 20		5 27 ·0			
Sun ☉ to E	36 9 0	+3 40	2 6 36 ·5	1	....	
	36 22 40		7 6 ·0			
	36 35 40		7 33 ·0			} h. m. s.
	36 48 0		8 0 ·0			C_{tm}= +5 15 31 ·9
	37 4 10		8 37 ·0			
Sun ☉ to E	43 5 20	+3 20	2 22 6 ·0	1	....	
	43 17 20		22 33 ·0			
	43 30 40		23 3 ·5			} h. m. s.
	43 47 0		23 38 ·0			C_{tm}= +5 15 29 ·9
	43 56 20		24 0 ·0			
Sun ☉ to E	44 26 0	+3 20	2 25 5 ·0	1	....	
	44 38 40		25 33 ·0			
	45 0 0		26 18 ·0			
	45 21 0		27 4 ·0			h. m. s.
	45 41 50		27 50 ·0			} C_{tm}= +5 15 32 ·0
	45 58 40		28 26 ·0			
	46 22 0		29 20 ·0			

Appendix B.

Civil Date, 1906.	Numerical Order.	Observer.	Place.	Latitude N.	Longitude E.G.	Barom. mm.	Thermom. annexed (Celsius).	Temp. of the Air (Celsius).	Position Lid of the Artificial Horizon.
				° ′ ″	h. m. s.		°	°	
29th June	41	Com. Cagni	Bujongolo	0 20 23	....	489 ·7	5 ·6	5 ·6	0A △
,,	42	,,	,,	,,	....	,,	,,	,,	0A ▽
9th July	43	S.A.R.	,,	....	2 0 6	489 ·5	13 ·0	11 ·0	0A
10th July	44	Com. Cagni	,·	....	,,	488 ·5	8 ·2	9 ·0	0A
,,	45	,,	,,	....	....	....	....	....	....
,,	46	,,	,,	....	2 0 6	488 ·5	8 ·2	9 ·0	0A
11th July	47	H.R.H.	,,	0 20 23	....	489 ·7	6 ·0	6 ·0	0A △
,,	48	,,	,,	,,	....	,,	,,	,,	0A ▽
,,	49	,,	,,	....	2 0 6	488 ·5	7 ·0	6 ·0	0A
,,	50	,,	,,	....	,,	,,	,,	,,	,,
,,	51	,,	,,	....	,,	,,	,,	,,	,,
,,	52	,,	,,	....	,,	,,	,,	,,	,,
,,	53	,,	,,	....	,,	,,	,,	,,	,,
,,	54	,,	,,	....	,,	,,	,,	,,	,,

I.—Astronomic Observations.

Conditions under which the Aster was observed.	Instrumental Height.	Instrument of Correction.	Hour of the Chronometer.	Number of the Chronometer.	Absolute Correction of the Chronometer.	Results.
	° ′ ″	′ ″	h. m. s.		h. m. s.	
Sun ☉ to E	62 46 40	+ 1 52·5	3 6 12·0	1	....	h. m. s.
	63 3 40		6 50·0			$C_{tm} = +5\ 15\ 26\cdot8$
	63 17 0		7 19·0			
	63 36 0		8 3·0			
	63 57 0		8 50·5			
Sun ☉ to E	64 37 0	+ 1 52·5	3 10 19·5	1	....	h. m. s.
	64 53 0		10 53·5			
	65 17 40		11 52·0			$C_{tm} = +5\ 15\ 29\cdot4$
	65 36 0		12 28·0			
	65 47 0		12 58·0			
Sun ☉ PS	136 17 40	+ 1 0	....	....	....	$\phi = 0° 20' 55'' $ N.
Sun ☉ Cir-cumm.	135 9 40	+ 1 0	6 39 5·0	1	+ 3 13 26·0	$\phi = 0° 20' 01'' $ N.
....	....	....	....	....	....	....
Sun ☉ Cir-cumm.	135 22 20	+ 1 0	6 44 24·0	1	+ 3 15 26·0	$\phi = 0° 19' 43'' $ N.
Sun ☉ to E	87 0 0	+ 1 50	3 49 42·0	2	....	h. m. s.
	87 20 0		50 26·0			
	87 40 0		51 14·0			$C_{tm} = +5\ 15\ 33\cdot0$
Sun ☉ to E	89 40 0	+ 1 50	3 55 58·0	2	....	h. m. s.
	90 0 0		56 44·0			
	90 20 0		57 31·3			$C_{tm} = +5\ 15\ 32\cdot3$
Sun ☉ Cir-cumm.	136 8 40	+ 1 50	6 34 14·5	1	....	$\phi = 0° 21' 12'' $ N.
Sun ☉ Cir-cumm.	136 42 0	+ 1 50	6 44 34·5	1	....	$\phi = 0° 20' 58'' $ N.
Sun ☉ Cir-cumm.	136 46 0	+ 1 50	6 51 29·0	1	....	$\phi = 0° 21' 13'' $ N.
Sun ☉ Cir-cumm.	136 43 0	+ 1 50	6 53 38·5	1	....	$\phi = 0° 20' 43'' $ N.
Sun ☉ Cir-cumm.	136 42 20	+ 1 50	6 54 12·0	1	....	$\phi = 0° 20' 46'' $ N.
Sun ☉ Cir-cumm.	136 41 0	+ 1 50	6 54 44·5	1	....	$\phi = 0° 20' 32'' $ N.

Appendix B.

Civil Date, 1906.	Numerical Order.	Observer.	Place.	Latitude N.	Longitude E.G.	Barom. mm.	Thermom. annexed (Celsius).	Temp. of the Air (Celsius).	Position Lid of the Artificial Horizon.
12th July	55	Com. Cagni	Bujongolo	° ′ ″ 0 20 23	h. m. s.	....	°	°	....
,,	56	H.R.H.	,,	,,	2 0 6	489·2	9·0	9·0	0A △
,,	57	,,	,,	,,	,,	,,	,,	,,	0A △
,,	58	,,	,,	,,	,,	,,	,,	,,	0A △
,,	59	,,	,,	,,	,,	,,	,,	,,	0A △
,,	60	Com. Cagni		,,	,,	489·0	7·0	7·0	0A
14th July	61	,,	Nakitawa	0 20 20	....	562·85	17·0	15·0	0A
16th July	62	,,	Ibanda	0 19 59	....	650·0	22·0	22·0	0A
17th July	63	,,	,,	,,	....	652·0	22·0	24·0	0A △
,,	64	,,	,,	,,	....	,,	,,	,,	0A ▽

I.—Astronomic Observations.

Conditions under which the Aster was observed.	Instrumental Height.	Instrument of Correction.	Hour of the Chronometer.	Number of the Chronometer.	Absolute Correction of the Chronometer.	Results.
✱ BAG 81— Emersion from the lunar disk	° ′ ″	′ ″	h. m. s. 10 14 4·0	1	h. m. s. +5 15 32·2 on the mean local time	h. m. s. λ = 2 00 6·3 E.G.
Sun ☉ to E	52 40 0 53 0 0 53 20 0	+3 50	2 44 40·0 45 23·0 46 9·0	1	....	h. m. s. C_{tm} = 5 15 34·9
Sun ☉ to E	53 40 0 54 0 0 54 20 0	+3 50	2 46 53·5 47 37·0 48 32·0	1	....	h. m. s. C_{tm} = 5 15 33·7
Sun ☉ to E.	55 40 0 56 0 0 56 20 0	+4 15	2 51 20·5 52 4·5 52 49·0	1	....	h. m. s. C_{tm} = +5 15 32·4
Sun ☉ to E.	57 0 0 57 20 0 57 40 0	+4 15	2 54 18·0 55 1·0 55 44·5	1	....	h. m. s. C_{tm} = +5 15 32·9
Sun ☉ to E.	56 56 0 55 12 0 55 28 0 55 40 0 55 55 0 56 15 0	+2 45	2 49 42·0 50 16·0 50 52·0 51 20·0 51 53·5 52 36·0	1	....	h. m. s. C_{tm} = +5 15 34·4
Sun ō to W.	65 4 40 64 8 0 63 5 0	+1 25	10 30 2·0 32 7·0 34 26·0	1	+3 15 26·4	h. m. s. λ = 2 0 19·6 E.G.
Sun ō to W.	27 23 0 27 5 0 26 41 0 26 17 40 25 52 40	+1 50	11 52 31·0 53 9·0 54 2·0 54 54·0 55 46·5	1	+3 15 23·3	h. m. s. λ = 2 0 43·9 E.G.
Sun ☉ to E.	44 51 0 45 9 0 45 27 40 45 46 40 45 59 20	+1 50	2 26 58·0 27 37·0 28 18·5 29 0·5 29 28·5	1	+3 15 22·4	h. m. s. λ = 2 0 41·5 E.G.
Sun ☉ to E.	46 23 20 46 42 0 46 58 0 47 8 40 47 30 0	+1 40	30 19·0 30 59·0 31 34·5 31 59·5 32 43·5	1	+3 15 22·4	h. m. s. λ = 2 0 42·5 E.G.

Appendix B.

Civil Date, 1906.	Numerical Order.	Observer.	Place.	Latitude N.	Longitude E.G.	Barom. mm.	Thermom. annexed (Celsius).	Temp. of the Air (Celsius).	Position Lid of the Artificial Horizon.
				° ′ ″	h. m. s.		°	°	
17th July	65	Com. Cagni	Ibanda	....	2 0 43	651·6	26·0	29·0	0A
18th July	66	,,	,,	....	,,	652·0	23·0	27·0	0A
19th July	67	,,	,,	....	,,	650·95	25·0	27·0	0A
,,	68	H.R.H.	.,	0 19 59	....	649·45	24·0	23·0	0A △
,,	69	,,	,,	,,	....	,,	,,	,,	0A ▽
20th July	70	,,	Butanuka	....	2 1 4	645·0	24·0	24·0	0A
,,	71	,,	,,	0 26 33	....	644·05	24·0	24·0	0A △
,,	72	,,	,,	,,	....	,,	,,	,,	0A ▽
21st July	73	,,	Fort Portal	0 39 28	2 1 32	638·6	21·0	21·0	0A △
,,	74	,,	,,	,,	,·	,,	:,	,,	0A △
22nd July	75	,,	,,	....	,,	641·0	19·0	20·0	0A
,,	76	,,	,,	0 39 28	,,	636·0	14·0	12·8	0A △
,,	77	,,	,,	,,	,·	:,	,,	,,	0A ▽
23rd July	78	,,	,,	,,	,,	641·7	17·5	17·0	0A △
,,	79	,,	,,	,,	,,	,,	,,	,·	0A ▽
24th July	80	,,	,,	,,	,,	640·1	18·0	16·5	0A △

I.—Astronomic Observations.

Conditions under which the Aster was observed.	Instrumental Height.	Instrument of Correction.	Hour of the Chronometer.	Number of the Chronometer.	Absolute Correction of the Chronometer.	Results.
	° ′ ″	′ ″	h. m. s.		h. m. s.	
Sun <u>O</u> PS	137 27 0	+ 2 55	....	....	....	φ = 0° 20′ 37″ N.
Sun <u>O</u> PS	137 47 30	+ 1 30	....	....	....	φ = 0° 20′ 4″ N.
Sun <u>O</u> PS	138 7 40	+ 1 37·5	....	....	....	φ = 0° 19′ 47″ N.
Sun <u>O</u> to W	55 40 0	+ 1 50	10 49 26·8	1	+ 3 15 18·8	h. m. s. λ = 2 0 42·7 E.G.
Sun <u>O</u> to W	52 20 0	+ 1 50	10 56 43·0	1	+ 3 15 18·8	h. m. s. λ = 2 0 43·5 E.G.
Sun O PS	139 46 0	+ 1 30				φ = 0° 26′ 33″ N.
Sun <u>O</u> to W	51 22 40 51 7 40 50 50 20	+ 2 10	10 58 52·0 10 59 25·0 11 0 1·5	1	+ 3 15 17·3	h. m. s. λ = 2 1 4·1 E.G.
Sun <u>O</u> to W	50 35 40 50 24 0 50 3 0	+ 2 10	11 0 34·0 0 59·5 1 44·5	1	+ 3 15 17·3	h. m. s. λ = 2 1 4·6 E.G.
Sun <u>O</u> to W	54 11 40 53 43 0 53 52 40 53 7 20	+ 1 20	10 52 54·5 53 55·0 54 32·0 55 13·0	1	....	h. m. s. κ_1 = + 3 15 16·1
Sun <u>O</u> to W	52 46 40 47 39 40	+ 1 20	10 55 57·5 11 7 40	1	....	h. m. s. κ_1 = + 3 15 15·9
Sun O PS	140 57 20	+ 2 20				φ = 0° 39′ 59″ N.
Sun <u>O</u> to E	49 36 40 49 53 20 50 4 20	+ 1 20	2 35 44·0 36 20·0 36 44·5	1	....	h. m. s. κ_1 = + 3 15 17·5
Sun <u>O</u> to E	50 26 0 50 38 40 50 51 20	+ 1 20	2 37 32·5 38 0·0 38 26·5	1	....	h. m. s. κ_1 = + 3 15 16·5
Sun <u>O</u> to E	49 20 0 49 40 0 50 0 0	+ 1 30	2 35 5·5 35 44·0 36 26·0	1	....	h. m. s. κ_1 = + 3 15 18·8
Sun <u>O</u> to E	50 20 0 50 40 0 51 0 0	+ 1 30	2 37 11·5 37 53·5 38 36·5	1	....	h. m. s. κ_1 = + 3 15 20·1
Sun <u>O</u> to E	55 40 0 56 0 0 56 20 0	+ 1 50	2 48 35·5 49 19·0 50 0·0	1	....	h. m. s. κ_1 = 3 15 22·6

Appendix B.

Civil Date, 1906.	Numerical Order.	Observer.	Place.	Latitude N.	Longitude E.G.	Barom. mm.	Thermom. an-nexed (Celsius).	Temp. of the Air (Celsius).	Position Lid of the Artificial Horizon.
				° ′ ″	h. m. s.		°	°	
24th July	81	H.R.H.	Fort Portal	0 39 28	2 1 32	640 ·1	18 ·0	16 ·5	OA ▽
27th July	82	,,	,,	,,	,,	635 ·0	17 ·0	16 ·0	OA △
,,	83	,,	,,	,,	,,	,,	,,	,,	OA ▽
28th July	84	,,	,,	,,	,,	636 ·0	16 ·0	14 ·0	OA △
,,	85	,,	,,	,,	,,	,,	,,	,,	OA ▽
,,	86	,,	,,	....	,,	635 ·0	24 ·0	23 ·0	OA −
31st July	87	,,	,,	0 39 28	,,	636 ·0	15 ·6	15 ·6	OA △
,,	88	,,	,,	,,	,,	,,	,,	,,	OA ▽
1st August	89	,,	Butiti	0 39 30	....	645 ·7	19 ·0	18 ·0	OA △
,,	90	,,	,,	,,	....	,,	,,	,,	OA ▽
2nd August	91	,,	Kaibo	....	2 3 8	652 ·0	24 ·0	24 ·0	OA −
,,	92	,,	,,	....	,,	,,	,,	,,	,,
,,	93	,,	,,	....	,,	,,	,,	,,	,,
,,	94	,,	,,	....	,,	,,	,,	,,	,,

* The Chronometer

336

I.—Astronomic Observations.

Conditions under which the Aster was observed.	Instrumental Height.	Instrument of Correction.	Hour of the Chronometer.	Number of the Chronometer.	Absolute Correction of the Chronometer.	Results.
	° ′ ″	′ ″	h. m. s.		h. m. s.	
Sun ☉ to E	56 40 0 57 0 0 57 20 0	+ 1 50	2 50 46 ·5 51 27 ·5 52 12 ·0	1	....	h. m. s. }κ_1 = + 3 15 22 ·6
Sun ☉ to W	62 0 0 61 40 0 61 20 0	+ 0 50	10 19 26 ·0 20 9 ·0 20 58 ·5	1	....	h. m. s. }κ_1 = + 3 32 57 ·6*
Sun ☉ to W	60 40 0 60 20 0 60 0 0	+ 0 50	10 22 9 ·5 23 2 ·0 23 46 ·5	1	....	h. m. s. }κ_1 = + 3 32 51 ·0
Sun ☉ to E	51 0 0 51 20 0 51 40 0	+ 0 30	2 20 17 ·5 20 59 ·0 21 42 ·0	1	....	h. m. s. }κ_1 = + 3 32 59 ·5
Sun ☉ to E	52 0 0 52 20 0 52 40 0	+ 0 30	2 22 25 ·0 23 7 ·5 23 51 ·5	1	....	h. m. s. }κ_1 = + 3 32 59 ·9
Sun ☉ PS	142 25 40	+ 2 20	....	....	....	$\phi = 0° 39′ 40″$ N.
Sun ☉ to W	59 40 0 59 20 0 59 0 0	+ 1 30	10 25 2 ·5 25 43 ·5 26 26 ·0	1	....	h. m. s. }κ_1 = + 3 33 4 ·8
Sun ☉ to W	58 40 0 58 20 0 58 0 0	+ 1 30	10 27 11 ·5 27 52 ·0 28 35 ·0	1	....	h. m. s. }κ_1 = + 3 33 4 ·7
Sun ☉ to W	52 20 0 52 0 0 51 20 0 51 0 0	+ 0 40	10 39 46 ·5 40 32 ·0 41 55 ·5 42 38 ·0	1	+ 3 33 6 ·5	h. m. s. }λ = 2 2 34 ·3 E.G.
Sun ☉ to W	50 40 0 50 20 0 50 0 0	+ 0 40	10 43 21 ·0 44 2 ·5 44 45 ·5	1	+ 3 33 6 ·5	h. m. s. }λ = 2 2 34 ·8 E.G.
Sun ☉ Circumm.	143 1 40	+ 1 30	6 51 4 ·0	1	+ 3 33 7 ·8	$\phi = 0° 30′ 36″$ N.
Sun ☉ Circumm.	142 51 20	,,	6 52 10 ·0	1	+ 3 33 7 ·8	$\phi = 0° 30′ 9″$ N.
Sun ☉ Circumm.	142 31 40	,,	6 54 9 ·5	1	+ 3 33 7 ·8	$\phi = 0° 29′ 26″$ N.
Sun ☉ Circumm.	142 21 0	,,	6 55 25 ·0	1	+ 3 33 7 ·8	$\phi = 0° 30′ 14″$ N.

remained unregulated.

Appendix B.

Civil Date, 1906.	Numerical Order.	Observer.	Place.	Latitude N.	Longitude E.G.	Barom. mm.	Thermom. annexed (Celsius).	Temp. of the Air (Celsius).	Position Lid of the Artificial Horizon.
				° ′ ″	h. m. s.		°	°	
2nd August	95	H.R.H.	Kaibo	0 30 4	....	651 ·0	23 ·0	23 ·0	0A △
,,	96	,,	,,	,,	....	,,	,,	,,	0A ▽
6th August	97	,,	Muyongo	0 30 41	....	657 ·0	25 ·0	25 ·0	0A △
,,	98	,,	,,	,,	....	,,	,,	,,	0A ▽
,,	99	,,	Kichiomi	....	2 4 27	653 ·0	24 ·0	24 ·0	0A
,,	100	,,	,,	0 31 20	....	652 ·0	22 ·0	22 ·0	0A △
,,	101	,,	,,	,,	....	,,	,,	,,	0A ▽
7th August	102	,,	Lwamutu-kuza	....	2 5 16	657 ·0	22 ·0	23 ·0	0A
8th August	103	,,	Kasiba	....	2 5 52	658 ·0	23 ·0	23 ·0	0A
,,	104	,,	,,	0 40 34	....	657 ·0	24 ·0	24 ·0	0A △
,,	105	,,	,,	,,	....	,,	,,	,,	0A ▽
9th August	106	,,	Kijemula	....	2 6 0	661 ·0	26 ·0	26 ·0	0A
10th August	107	,,	Bimbye	....	2 7 26	662 ·0	27 ·0	26 ·0	0A
11th August	108	,,	,,	0 31 56	....	664 ·0	26 ·0	26 ·5	0A △
,,	109	,,	,,	,,	....	,,	,,	,,	0A ▽

* The heights marked with an asterisk were taken with the Magnaghi circle and

I.—Astronomic Observations.

Conditions under which the Aster was observed.	Instrumental Height.	Instrument of Correction.	Hour of the Chronometer.	Number of the Chronometer.	Absolute Correction of the Chronometer.	Results.
	° ′ ″	′ ″	h. m. s.		h. m. s.	
Sun ☉ to W.	51 20 0 51 0 0 50 40 0	+ 2 10	10 41 9·0 41 52·5 42 32·5	1	+ 3 33 8·0	h. m. s. }λ = 2 3 7·8 E.G.
Sun ☉ to W.	50 20 0 50 0 0 49 49 0	+ 2 10	10 43 16·0 43 58·5 44 41·5	1	+ 3 33 8·0	h. m. s. }λ = 2 3 7·9 E.G.
Sun ☉ to W.	20 58 20 20 37 40 20 21 40	+ 2 15	11 44 33·0 45 18·0 45 52·0	1	+ 3 33 12·6	h. m. s. }λ = 2 3 55·4 E.G.
Sun ☉ to W.	19 54 40	+ 2 15	11 46 48·0	1	+ 3 33 12·6	h. m. s. λ = 2 3 54·6 E.G.
Sun ☉ PS	146 50 30*	− 8 0	....	....	....	φ = 0° 31′ 53″ N.
Sun ☉ to W.	45 0 0 44 40 0 44 20 0	+ 0 40	10 53 31·0 54 12·0 54 52·5	1	+ 3 33 14·1	h. m. s. }λ = 2 4 26·0 E.G.
Sun ☉ to W.	44 0 0 43 40 0 43 20 0	+ 0 40	10 55 36·0 56 18·0 57 1·0	1	+ 3 33 14·1	h. m. s. }λ = 2 4 25·9 E.G.
Sun ☉ PS	147 13 30*	+ 1 0	....	....	....	φ = 0° 30′ 27″ N.
Sun ☉ PS	148 6 0*	0 0	....	....	....	φ = 0° 40′ 34″ N.
Sun ☉ to W	54 0 0 53 40 0 53 20 0	+ 1 50	10 33 16·0 33 56·0 34 40·5	1	+ 3 33 17·1	h. m. s. }λ = 2 5 53·2 E.G
Sun ☉ to W	53 00 0 52 40 0 52 20 0	+ 1 50	10 35 23·0 36 6·5 36 49·5	1	+ 3 33 17·1	h. m. s. }λ = 2 5 50·8 E.G.
Sun ☉ PS	148 30 45*	0 0	....	....	....	φ = 0° 35′ 55″ N.
Sun ☉ PS	148 55 45*	+ 1 30	....	....	....	φ = 0° 31′ 57″ N.
Sun ☉ to W	51 0 0 50 40 0 50 20 0	+ 1 20	10 37 49·5 38 31·0 32 13·0	1	+ 3 33 21·6	h. m. s. }λ = 2 7 26·4 E.G.
Sun ☉ to W	50 0 0 49 40 0 49 20 0	+ 1 20	10 39 55·0 40 37·0 41 17·5	1	+ 3 33 21·6	h. m. s. }λ = 2 7 26·4 E.G.

result from the mean of the readings of the two reflectors of the instrument.

Appendix B.

Civil Date, 1906.	Numerical Order.	Observer.	Place.	Latitude N.	Longitude E.G.	Barom. mm.	Thermom. annexed (Celsius).	Temp. of the Air (Celsius).	Position Lid of the Artificial Horizon.
16th August	110	H.R.H.	Entebbe	° ′ ″ 0 3 11	h. m. s. 2 9 47	667 ·5	° 22 ·0	° 24 ·0	0A △
,,	111	,,	,,	,,	,,	,,	,,	,,	0A ▽
17th August	112	,,	,,	·,	,,	666 ·0	17 ·0	24 ·0	0A △
,,	113	,,	,,	,,	,,	,,	,,	,,	0A ▽

I.—Astronomic Observations.

Conditions under which the Aster was observed.	Instrumental Height.	Instrument of Correction.	Hour of the Chronometer.	Number of the Chronometer.	Absolute Correction of the Chronometer.	Results.
	° ′ ″	′ ″	h. m. s.			h. m. s.
Sun ☉ to **W**	54 0 0 53 40 0 53 20 0	+2 0	10 28 25 ·5 29 6 ·5 29 48 ·0	1	....	} $\kappa_1 = +3\ 33\ 29$ ·9
Sun ☉ to **W**	53 0 0 52 40 0 52 20 0	+2 0	10 30 30 ·0 31 13 ·0 31 55 ·5	1	....	h. m. s. } $\kappa_1 = +3\ 33\ 28$ ·5
Sun ☉ to **E**	68 20 0 68 40 0 69 0 0	+1 50	2 43 13 ·0 43 55 ·0 44 36 ·2	1	....	h. m. s. } $\kappa_1 = +3\ 33\ 25$ ·2
Sun ☉ to **E**	69 20 0 69 40 0 70 0 0	+1 50	2 45 20 ·5 46 0 ·5 46 43 ·5	1	....	h. m. s. } $\kappa_1 = +3\ 33\ 23$ ·8

REGISTER
OF THE CHRONOMETERS.

Appendix B.

Civil Date, 1906.	Place.	Temperature.	Approximate Time.	Chronom. N. 1, Lange 56509.		
				C₁.	K₁.	k₁.
			h. m.	h. m. s.		
18th April	On board		12 30	9 15 0·0		
19th ,, ...	Port Said		9 30	6 0 0·0		
20th ,, ...	,, ...		9 30	6 21 0·0	h. m. s.	
20th ,, ...	Port Said (comparison with the chron. of the Police Station)		14 0	10 46 8·5	+ 3 17 15·5	
20th ,, ...	Port Said		6 0	2 57 0·0		
21st ,, ...	On board	°	10 0	6 33 0·0		s.
22nd ,, ...	,, ...	25	10 0	6 55 0·0		−1 ·29
23rd ,, ...	,, ...	28	10 0	6 36 0·0		
24th ,, ...	,, ...	29	9 30	6 10 0·0		
25th ,, ...	,, ...	30	9 0	5 55 0·0		
26th ,, ...	Jibuti (comparison with the " Elphinstone ")		16 0	12 40 30·0	+ 3 17 8·0	
26th ,, ...	Jibuti	31	7 30	3 59 30·0		
27th ,, ...	Aden	29	9 0	5 43 0·0		
28th ,, ...	On board	28 ·5	9 0	5 54 0·0		
28th ,, ...	,, ...		10 30	7 6 30·0		
29th ,, ...	,, ...		9 0	5 26 0·0		s.
30th ,, ...	Mombasa		9 0	5 53 30·0		−1 ·34
1st May	,, ...	30	9 0	5 44 0·0		
2nd ,, ...	,, ...	31	9 30	6 19 0·0		
3rd ,, ...	,, ...		8 0	4 47 0·0		
4th ,, ...	,, ...		5 30	2 12 0·0		
4th ,, ...	,, (comparison with the chron. of the Post Office)				+ 3 16 57 ·0	

I.—Astronomic Observations.

THE CHRONOMETERS.

Chronom. N. 2, Lange 56520.			Chronom. N. 3, Longines 560229.			Comparisons.		
C_2.	K_2.	k_2.	C_3.	K_3.	k_3.	C_1-C_2.	C_1-C_3.	C_2-C_3.
h. m. s. 9 7 39·5			h. m. s. 9 32 23·5			m. s. + 7 20·5	m. s. −17 23·5	m. s. −24 44·0
5 52 38·5			6 17 19·0			7 21·5	17 19·0	24 40·5
6 13 37·0			6 38 15·0			7 23·0	17 15·0	24 38·0
	h. m. s. +3 24 38·5			h. m. s. +3 0 0·5				
2 49 36·2			3 14 14·5			+ 7 23·8	−17 14·5	−24 38·3
6 25 34·0			6 50 10·5			7 26·0	17 10·5	24 36·5
6 47 30·0		s. +4·31	7 12 0·7		s. +6·46	7 30·0	17 0·7	24 30·7
6 28 25·2			6 52 53·5			7 34·8	16 53·5	24 28·3
6 2 19·0			6 26 46·5			7 41·0	16 46·5	24 27·5
5 47 11·0			6 11 38·0			7 49·0	16 38·0	24 27·0
	+3 25 3·5			+3 0 38·0				
3 51 34·5			4 16 0·0			+ 7 55·5	−16 30·0	−24 25·5
5 34 57·5			5 59 20·5			8 2·5	16 20·5	24 23·0
5 45 49·5			6 10 4·0			8 10·5	16 4·0	24 14·5
6 58 19·0			7 22 33·7			8 11·0	16 3·7	24 14·7
5 17 41·5		s. +5·90	5 41 54·0		s. +7·91	8 18·5	15 54·0	24 12·5
5 45 4·5			6 9 15·0			8 25·5	15 45·0	24 10·5
5 35 26·0			5 59 38·0			8 34·0	15 38·0	24 12·0
6 10 18·0			6 34 30·0			8 42·0	15 30·0	24 12·0
4 38 6·5			5 2 21·5			8 53·5	15 21·5	24 15·0
2 3 6·0			2 27 16·8			8 54·0	15 16·8	24 10·8
	+3 25 52 0			+3 1 43·0				

Appendix B.

Civil Date, 1906.	Place.	Temperature.	Approximate Time.	Chronom. N. 1, Lange 56509.		
				C_1.	K_1.	k_1.
			h. m.	h. m. s.		
4th May	Mombasa		8 30	5 20 30 ·0		
5th ,, ...	Railway		10 0	6 49 0 ·0		
6th ,, ...	,, ...		8 30	4 59 0 ·0		
7th ,, ...	Entebbe		9 30	6 22 0 ·0		
8th ,, ...	,, ...		9 30	6 19 0 ·0		s. −2 ·762
9th ,, ...	,, ...		10 30	7 14 0 ·0		
10th ,, ...	,, ...		10 0	6 45 0 ·0		
11th ,, ...	,, ...		10 30	7 10 0 ·0		
12th ,, ...	,, ...		8 32	5 15 0 ·0	h. m. s. +3 16 34 ·9	
12th ,,	Entebbe (Telegraphic comparison with Mombasa Post Office)		9 30			
12th ,, ...	Entebbe		10 30	7 9 0 ·0		
12th ,, ...	,, ...		10 30	7 11 0 ·0		
13th ,, ...	,, ...		13 30	10 18 0 ·0		
14th ,, ...	,, ...		11 0	7 40 0 ·0		
16th ,, ...	En route (Kutende)		10 0	6 44 0 ·0		
17th ,, ...	,, (Bweya)	29 ·2	9 30	6 16 0 ·0		
18th ,, ...	,, (Mitiana)....	28 ·7	10 0	6 31 0 ·0		
19th ,, ...	,, (Bujongolo)	30 ·7	9 30	6 10 30 ·0		
20th ,, ...	,, (Bimbye)....	27 ·7	9 30	6 26 0 ·0		
21st ,, ...	,, (Kijemula)	25 ·7	13 30	10 15 30 ·0		
22nd ,, ...	,, (Madridu)	28 ·2	10 0	6 55 0 ·0		
23rd ,, ...	,, (Muduma)	23 ·2	10 30	7 1 30 ·0		s. −2 ·123
24th ,, ...	,, (Lwamutukuza)	28 ·7	10 30	7 2 30 ·0		
25th ,, ...	,, (Kichiomi)	24 ·2	10 0	6 47 0 ·0		
26th ,, ...	,, (Muyongo)	26 ·7	10 0	6 55 0 ·0		
27th ,, ...	,, (Kaibo)	23 ·3	10 0	6 50 0 ·0		

I.—Astronomic Observations.

Chronom. N. 2, Lange 56520.			Chronom. N. 3, Longines 560229.			Comparisons.		
C_2.	K_2.	k_2.	C_3.	K_3.	k_3.	C_1-C_2.	C_1-C_3.	C_2-C_3.
h. m. s.			h. m. s.			m. s.	m. s.	m. s.
5 11 35 ·0			5 35 43 ·0			+ 8 55 ·0	−15 13 ·0	−24 8 ·0
6 40 0 ·0			7 4 6 ·5			9 0 ·0	15 6 ·5	24 6 ·5
4 49 58 ·5			5 13 58 ·8			9 1 ·5	14 58 ·8	24 0 ·3
6 12 53 ·0			6 36 53 ·0			9 7 ·0	14 53 ·0	24 0 ·0
6 9 49 ·0		s. +2 ·125	6 33 48 ·0		s. +3 ·437	9 11 ·0	14 48 ·0	23 59 ·0
7 4 43 ·5			7 28 43 ·0			9 16 ·5	14 43 ·0	23 59 ·5
6 35 37 ·5			6 59 38 ·0			9 22 ·5	14 38 ·0	24 0 ·5
7 0 31 ·5			7 24 31 ·5			+ 9 28 ·5	−14 31 ·5	−24 0 ·0
			5 29 25 ·3				14 25 ·3	
	h. m. s. +3 26 9 ·0		6 19 4 ·5	h. m. s. +3 2 10 ·5				
			7 23 23 ·5				−14 23 ·5	
7 1 25 ·5			7 25 24 ·0			+ 9 34 ·5	14 24 ·0	−23 58 ·5
10 8 20 ·5			10 32 18 ·0			9 39 ·5	14 18 ·0	23 57 ·5
7 30 16 ·0			7 54 13 ·5			9 44 ·0	14 13 ·5	23 57 ·5
6 33 56 ·0			6 57 55 ·5			10 4 ·0	13 55 ·5	23 59 ·5
6 5 47 ·5			6 29 39 ·5			10 12 ·5	13 39 ·5	23 52 ·0
6 20 38 ·5			6 44 30 ·5			10 21 ·5	13 30 ·5	23 52 ·0
5 59 59 ·5			6 24 1 ·0			10 30 ·5	13 31 ·0	24 1 ·5
6 15 18 ·5			6 39 26 ·5			10 41 ·5	13 26 ·5	24 8 ·0
10 4 41 ·0			10 28 50 ·0			10 49 ·0	13 20 ·0	24 9 ·0
6 44 0 ·0			7 8 10 ·0			11 0 ·0	13 10 ·0	24 10 ·0
6 50 21 ·0			7 14 23 ·5			11 9 ·0	12 53 ·5	24 2 ·5
6 51 10 ·0		s. +6 ·264	7 15 25 ·0		s. +5 ·087	11 20 ·0	12 55 ·0	24 15 ·0
6 35 31 ·5			6 59 51 ·0			11 28 ·5	12 51 ·0	24 19 ·5
6 43 19 ·0			7 7 46 ·0			11 41 ·0	12 46 ·0	24 27 ·0
6 38 9 ·0			7 2 39 ·0			11 51 ·0	12 39 ·0	24 30 ·0

347

Civil Date, 1906.	Place.					Temperature.	Approximate Time.	Chronom. **N.** 1, Lange 56509.		
								C₁.	K₁.	k₁.
						°	h. m.	h. m. s.		
28th May	En route	(Butiti)	....		....	22 ·2	11 0	7 39 0 ·0		
29th ,, ...	,,	(Fort Portal)			....	19 ·7	14 30	11 13 0 ·0		
30th ,, ...		,,			....	21 ·2	12 30	9 24 0 ·0		
31st ,, ...		,,			....	20 ·7	10 30	7 8 0 ·0		
1st June	,,	(Duwona)			....	23 ·7	10 0	6 58 0 ·0		
2nd ,, ...	,,	(Kasongo)			....	26 ·7	11 30	8 27 0 ·0		
3rd ,, ...	,,	(Ibanda)	....		....	25 ·2	10 30	7 11 0 ·0		
4th ,, ...	,,	(Bihunga)			....	20 ·7	12 30	9 26 0 ·0		
5th ,, ...	,,	(Nakitawa)			....	13 ·7	10 0	6 47 0 ·0		
11th ,, ...	Bujongolo	....		....	....	5 ·9	12 0	8 31 0 ·0		
11th ,, ...	,,		....	....	....		13 30	10 29 2 ·0	h. m. s. +3 15 30 ·8	
12th ,, ...	,,		....	....	....	4 ·9	10 0	6 56 0 ·0		
14th ,, ...	,,		....	....	....	4 ·9	14 0	10 43 0 ·0		
16th ,, ...	,,		....	....	....	5 ·9	7 30	4 10 0 ·0		
18th ,, ...	,,		....	....	....	4 ·9	8 30	5 24 0 ·0		
24th ,, ...	,,		....	....	....	4 ·9	8 30	5 20 30 ·0		s. −0 ·509
25th ,, ...	,,		....	....	....	5 ·2	9 30	6 21 0 ·0		
26th ,, ...	,,		....	....	....	4 ·7	10 30	7 6 0 ·0		
27th ,, ...	,,		....	...	....	6 ·5	10 0	6 55 30 ·0		
28th ,, ...	,,		....	....	....	4 ·7	10 0	6 51 0 ·0		
29th ,, ...	,,		....	....	....	4 ·7	6 30	3 9 34 ·0	+3 15 21 ·8	
29th ,, ...	,,		....	....	....		10 30	7 8 0 ·0		
30th ,, ...	,,		....	...	....	4 ·7	11 0	7 48 0 ·0		
1st July	,,		....	....	....	4 ·3	10 0	6 35 0 ·0		s. +0 ·471
2nd ,, ...	,,		....	....	....	6 ·0	10 0	6 50 0 ·0		
5th ,, ...	,,		....	....	....	4 ·5	10 30	7 26 0 ·0		

* Taking the comparisons of June 11th to be wrong, we are

I.—Astronomic Observations.

Chronom. N. 2, Lange 56520.			Chronom. N. 3, Longines 560229.			Comparisons.		
C_2.	K_2.	k_2.	C_3.	K_3.	k_3.	C_1-C_2.	C_1-C_3.	C_2-C_3.
h. m. s.			h. m. s.			m. s.	m. s.	m. s.
7 27 0·5			7 51 33·0			11 59·5	12 33·0	24 32·5
11 0 48·0			11 25 16·0			12 12·0	12 16·0	24 28·0
9 11 34·0			9 36 2·0			12 26·0	12 2·0	24 28·0
6 55 28·0			7 19 55·5			12 32·0	11 55·5	24 27·5
6 45 17·0			7 9 40·5			12 43·0	11 40·5	24 23·5
8 14 8·0			8 38 30·0			12 52·0	11 30·0	24 22·0
6 57 58·0			7 22 22·0			13 2·0	11 22·0	24 24·0
9 12 47·5			9 37 13·0			13 12·5	11 13·0	24 25·5
6 33 37·0			6 58 0·0			13 23·0	11 0·0	24 23·0
8 17 9·0			8 41 11·0			? *	?	?
	h. m. s. +3 29 23·3			h. m. s. +3 4 48·3				
6 42 7·0			7 6 42·0			+13 53·0	−10 42·0	−24 35·0
10 29 5·5			10 53 24·0			13 54·5	10 24·0	24 18·5
3 56 8·0			4 20 20·5			13 52·0	10 20·5	24 12·5
5 10 17·0			5 34 19·5			13 43·0	10 19·5	24 2·5
5 6 54·5		s. −2·611	5 30 4·0		s. +5·197	13 35·5	9 34·0	23 9·5
6 7 30·0			6 30 34·5			13 30·0	9 34·5	23 4·5
6 52 32·5			7 15 22·0			13 27·5	9 22·0	22 49·5
6 42 7·0			7 4 54·5			13 23·0	9 24·5	22 47·5
6 37 40·0			7 0 17·5			13 20·0	9 17·5	22 37·5
	+3 28 39·3			+3 6 15·9				
6 54 43·0			7 17 3·5			13 17·0	9 3·5	22 20·5
7 34 44·5			7 56 45·5			13 15·5	8 45·5	22 1·5
6 21 47·5			6 43 45·5			13 12·5	8 45·5	21 58·0
6 36 53·5			6 58 2·0			13 6·5	8 2·0	21 8·5
7 13 14·0			7 34 37·0			12 46·0	8 37·0	21 23·0

referred to those of the 12th for the values K_2, K_3 and k_2, k_3.

Appendix B.

Civil Date, 1906.	Place.	Temperature.	Approximate Time.	Chronom. N. 1, Lange 56509.		
				C_1.	K_1.	k_1.
		°	h. m.	h. m. s.	h. m. s.	
11th July	Bujongolo	5·9	8 0	4 46 28·5		
13th ,, ...	,, (en route)	5·9	6 0	2 51 6·0	+3 15 28·4	
						s. −1·521
21st ,, ...	Fort Portal		14 0	10 57 49·0	+3 15 15·8	
24th ,, ...	,, ...	14·2	6 30	2 13 0·0		

27th July—Through a great delay in regulating

Civil Date	Place.	Temperature.	Approximate Time.	C_1	K_1	k_1
			h. m.	h. m. s.	h. m. s.	
27th July ...	Fort Portal		14 0	10 32 0·0		
28th ,, ...	,, ...		9 0	5 40 0·0		
30th ,, ...	,, ...		10 0	6 29 0·0		
31st ,, ...	,, ...		14 0	10 26 48·0	+3 33 5·0	
2nd August	Kaibo		13 30	9 46 0·0		
6th ,,	Kichiomi		15 30	11 45 0·0		s. +1·510
16th ,,	Bimbye		10 0	6 40 0·0		
16th ,,	Entebbe		14 0		+3 33 29·2	
22nd ,,			10 0	6 37 0·0		s. −1·045
26th ,,	Mombasa (comparison with the chronometer of the Post Office)		7 30	4 4 0·0	+3 33 19·0	
28th ,,			14 30	10 45 0·0		

Chronom. N. 2, Lange 56520.			Chronom. N. 3. Longines 560229.			Comparisons.		
C_2.	K_2.	k_2.	C_3.	K_3.	k_3.	$C_1 - C_2$.	$C_1 - C_3$.	$C_2 - C_3$.
h. m. s. 7 34 0 ·0			h. m. s.			m. s. 12 28 ·5	m. s.	m. s.
1 59 50 ·0			2 19 40 ·5			13 10 ·0	6 40 ·5	19 50 ·5

the chronometers they varied, and No. 2 stopped.

h. m. s. 8 29 43 ·0			h. m. s. 10 17 47 ·5			h. m. s. +2 2 17 ·0	h. m. s. +0 14 12·5	h. m. s. −1 48 4·5
3 37 43 ·5			5 25 55 ·0			2 2 16 ·5	14 5·0	1 48 11·5
4 26 39 ·5			6 15 14 ·0			2 2 20 ·5	13 46·0	1 48 34·5
7 43 17 ·5			9 32 14 ·0			2 2 42 ·5	13 46·0	1 48 56·5
9 41 50 ·0						2 3 10 ·0		
4 36 13 ·0						2 3 47 ·0		
4 31 43 ·0						+2 5 17 ·0		
8 39 5 ·5						2 5 55·0		

II.—GEODETIC OBSERVATIONS.

By P. CAMPIGLI.

Taking as starting point a site near Bujongolo, the height of which above sea-level was known from barometric readings, at that point, which for shortness will henceforth be simply called Bujongolo, was constructed an astronomic station, the latitude of which was determined by meridian and circummeridian zenithal observations of the sun, and the longitude by means of lunar occultations of stars.

Then in the neighbourhood of Bujongolo a base was chosen and measured between the points A and B (*see* the annexed diagram of the triangulation.)

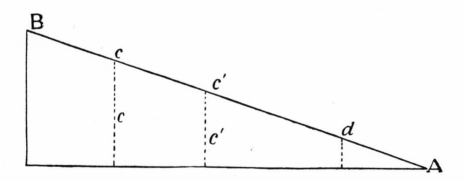

The distance was obtained either directly by fixing stations with the tachometer at the two extremes A and B, or by measuring with the metric

Appendix B.

tape measure and stadiometer, the distances intercepted between the points
B c, c c′, c′ d, d A, from which were had the following results:—

Distances.			Vertical Angle.			Segments.	
Between the points.	Measured with the tape line.	Measured with the Stadiometer.					
B c	56 ·135	60 ·00	15° 10′	55 ·89	}	} 55 ·905	
c B		61 ·5	17 32	55 ·92			
B A		320 ·0	18 02				289 ·33
c c′	67 ·620	75 ·0	21 20	65 ·07	} 165 ·98	} 166 ·375	
c′ d	108 ·325	118 ·0	22 22	100 ·91			
c d		190 ·5	20 40		166 ·77		
d A	62 ·700	66 ·5	7 06	65 ·48	}	} 65 ·235	
A d		66 ·0	7 12	64 ·99			288 ·45
A B		320 ·0	18 18			287 ·515	287 ·51
			Total length to be adopted, metres				288 ·43 (947 ft.)

At the extreme west point *B* of the base, the height of which over
Bujongolo was also measured with the barometer and found to be 104 metres,
solar observations enabled the expedition to determine the azimuth of one of
the points constituting the apexes of the triangles of the geodetic network,
which thus became orientated.

The point chosen for the azimuth was the Cagni Peak, distant 1882·9 metres
(5,980 feet), and the following results were obtained:—

> Mean of four values on the right
> > Circle..................80° 51′· 29.
>
> Mean of four values on the left
> > Circle..................80° 51′· 09.
>
> Mean value of the azimuth of Cagni Peak from Point B...80° 51′·19.

With the mean value of the base of 288·43 metres (945 feet), and solving
the two triangles *B A* Cagni Peak and *B A* Edward Peak, we obtained from
two parts the value of the side Cagni Peak—Edward Peak, which gave
2883·2 metres (9,456 feet).

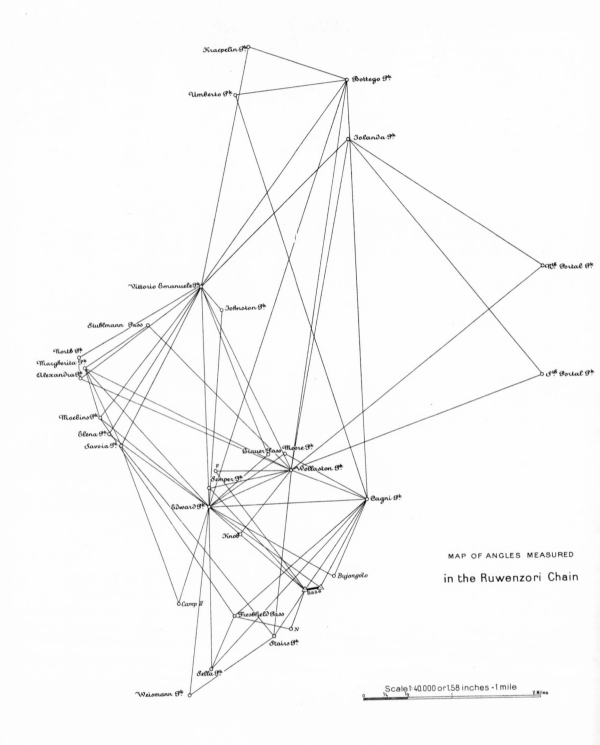

Kraepelin Pk.

Bottego Pk.

Umberto Pk.

Iolanda Pk.

Mt. Portal Pk.

Vittorio Emanuele Pk.

Johnston Pk.

Stuhlmann Pass

Sth Portal Pk.

North Pk.
Margherita Pk.
Alexandra Pk.

Moebius Pk.

Elena Pk.
Savoia Pk.

Grauer Pass   Moore Pk.

F
Semper Pk.   Wollaston Pk.

Edward Pk.   Cagni Pk.

Knob

Bujongolo

Base

MAP OF ANGLES MEASURED

in the Ruwenzori Chain

Camp II

Freshfield Pass

N

Stairs Pk.

Sella Pk.

Scale 1: 40.000 or 1,58 inches = 1 mile

0   ¼   ½   1   2 Miles

Weismann Pk.

II.—Geodetic Observations.

To this side were connected all the points, Bujongolo included, which formed the apexes of the network in which was comprised all the surveyed section of the Ruwenzori Range.

At all these points a station was made by measuring both the vertical and the horizontal angles by means of a prismatic compass from F. Barker and Son, London, No. 1926, except only for the two extreme points A and B of the base, where was employed the tachometer, and for the Cagni Peak, at which the observations were carried out with a small field theodolite.

In the tabulated Summary A are indicated all the triangles dealt with, as well as the value of the observed angles and that of the calculated sides.

The Summary B contains the orthogonal co-ordinates of all the points referred to the extreme west of the base B.

In the Summary C are brought together the heights of some points which are referred to Bujongolo, and were obtained by means of geodetic observations.

Lastly, Summary D is an epitome of the heights that were adopted for all the points indicated on the maps, as well as the method by which said numbers were deducted.

Appendix B.

Summary A.
LIST OF THE TRIANGLES.

Name of the Points.	Angles.	Sides, metres.	Name of the Points.	Angles.	Sides, metres.
Wollaston ...	157 37	2883·2	Johnston ...	133 00	3894·9
Edward	21 07	1485·0	Vittorio Eman.	41 00	3493·9
Cagni	23 16	1628·2	Edward ...	6 00	556·7
Margherita ...	21 30	1628·2	Knob	73 33·8	2098·1
Edward	109 30	4187·9	B	51 00·8	1709·4
Wollaston ...	49 00	3352·9	Wollaston ...	56 25·4	1832·2
Vittorio Emanuele	24 40	1628·2	Stuhlmann ...	102 06·2	4200·2 m.
Edward ...	68 40	3634·2	Margherita ...	58 53·8	3678·1
Wollaston ...	86 40	3894·9	Wollaston ...	19 00·0	1398·5
Margherita ...	59 00	3634·2	F	49 53·0	2098·1
Wollaston ...	57 30	2581·0	B	27 46·7	1278·3
Vittorio Emanuele	83 30	4212·5	Wollaston ...	102 20·3	2680·2
Iolanda	35 00	3634·2	Moore ...	117 59	2883·2
Vittorio Emanuele	108 00	6025·9	Edward ...	31 30	1705·9
Wollaston ...	37 00	3813·1	Cagni	30 31	1657·9
Bottego	10 00	1628·2	Weismann ...	48 58·0	2578·5
Edward	48 00	6968·2	Stairs ...	97 49·5	3386·4
Wollaston ...	122 00	7951·9	Edward ...	33 12·5	1872·1
Umberto	79 07	7429·7	Portal, North	71 12·5	6025·9
Cagni ...	15 37·5	2037·8	Iolanda ...	66 56·0	5856·3
Bottego	85 15·5	7539·9	Wollaston ...	41 50·5	4247·4
Alexandra ...	58 30	3634·2	Semper ...	104 00	1628·2
Wollaston ...	40 30	2768·1	Edward ...	65 00	1520·9
Vittorio Emanuele	81 00	4209·8	Wollaston ...	11 00	320·2
Margherita (N.Pk.)	98 00	2768·1	N	104 10·1	1527·6
Alexandra ...	74 00	2687·0	B	42 08·2	1057·0
Vittorio Emanuele	8 00	389·0	Freshfield ...	33 41·7	874·1
Kraepelin ...	85 30	4521·7	Camp II ...	39 00	1958·4
Vittorio Emanuele	24 30	1880·9	Edward ...	105 00	3005·9
Bottego	70 00	4262·2	Savoia ...	36 00	1829·1
Elena	90 30	3894·9	Moebius ...	87 30	3894·9
Edward	54 00	3151·2	Edward ...	51 30	3051·1
Vittorio Emanuele	35 30	2261·9	Vittorio Eman.	41 00	2557·7
Savoia	53 20	2579·8	Bujongolo ...	83 00·0	2883·2
Vittorio Emanuele	28 30	1534·7	Cagni ...	65 30·2	2643·2
Margherita ...	98 10	3183·6	Edward ...	31 29·8	1517·6
Sella	44 16	2883·2	Portal, South	70 12·5	6025·9
Cagni	44 14	2881·4	Iolanda ...	49 56·0	4901·1
Edward	91 30	4129·4	Wollaston ...	59 51·5	5538·2
Grauer Pass ...	107 30	3894·9			
Vittorio Emanuele	20 00	1396·8			
Edward	52 30	3240·0			

II.—Geodetic Observations.

ORTHOGONAL CO-ORDINATES OF THE POINTS, REFERRED TO B (EXTREME WEST BASE).

Name of the Points.	y	x
B—Extreme west base	0	0
A—Extreme east base	+ 284·8	+ 45·8
Cagni	+ 1028·3	− 1505·2
Edward	− 1854·5	+ 1458·2
Wollaston	− 345·4	+ 2069·5
Vittorio Emanuele	− 1903·4	+ 5352·8
Margherita	− 4086·7	+ 3978·5
Iolanda	+ 867·7	+ 4972·0
Umberto	− 1168·0	+ 8718·1
Alexandra	− 4187·8	+ 3789·5
Kraepelin	− 933·9	+ 9503·2
Elena	− 3700·9	+ 2764·7
Savoia	− 3462·0	+ 2576·8
Grauer Pass	− 757·2	+ 2322·3
Stuhlmann	− 2899·0	+ 4716·8
F	− 1622·5	+ 2133·3
Moore	− 455·3	+ 2347·4
Weismann	− 2311·1	− 1897·3
Stairs	− 746·1	− 869·9
Roccati	− 2805·6	+ 2714·8
Bottego	+ 858·5	+ 8933·0
Margherita, North Peak	− 4310·5	+ 4158·7
Knob	− 1594·5	+ 902·5
Portal, North	+ 4360·5	+ 5555·2
Portal, South	+ 4301·6	+ 3626·8
Camp II	− 2475·0	− 262·5
E	− 3876·0	+ 3025·2
Johnston	− 1532·9	+ 4937·3
Semper	− 1837·9	+ 1778·0
Sella	− 1882·9	− 1423·2
Bujongolo	+ 421·6	+ 114·2
N	− 420·5	− 766·3
Camp VII ... } Calculated {	+ 561·7	+ 7229·6
Lake Kujuku } to apex of {	− 1728·7	+ 2770·3
Freshfield Pass } pyramid {	− 1443·5	− 500·0

Appendix B.

RATIOS OF THE POINTS REFERRED TO BUJONGOLO, WHOSE HEIGHT ABOVE THE SEA IS 3,798 METRES (12,461 FEET).

The ratios of the points are calculated with the zenith distances observed at the various stations, and with the sides obtained from the triangulation.

Name of the Points.	Zenith Angles Observed.	Sides.	Differences of Level.	Ratios referred to Bujongolo.
Station B. Extreme west base, ratio determined with the barometer	..	..	..	104
Cagni Peak	+ 18° 47′	1822·9	+ 620	724
Edward „	+ 22 16	2359·1	+ 966	1070
Semper „	+ 18 27	2680·2	+ 894	998
Signal A	+ 11 31	288·4	− 94	10
Station A. Extreme east base, ratio ..	..	..	..	10
Cagni Peak	+ 23° 24′	1637·9	+ 709	719
Edward „	+ 22 30	2563·4	+ 1062	1072
Cagni Station, mean ratio	..	..	..	772
Edward	+ 6° 55′ 30″	2883·2	+ 350	1072
Margherita	+ 6 5 30	5681·6	+ 606	1328
Vittorio Emanuele	+ 4 31 30	4837·2	+ 383	1104
Umberto	+ 2 5 30	7539·9	+ 275	997
Iolanda	+ 1 39 30	8142·9	+ 235	957
Elena	+ 5 37 30	4894·0	+ 482	1204
Sella	+ 1 58 30	4129·2	+ 142	864
Wollaston	+ 5 33 30	1485·0	+ 144	866
Moore	+ 4 36 30	1705·9	+ 137	859
E	+ 4 42 30	5134·4	+ 423	1144
Stairs	+ 1 30 0	2961·5	+ 78	800
Vittorio Emanuele Station, mean ratio ..	..	..	..	1104
Margherita	+ 5° 0′	2581·0	+ 226	1330
Savoia	+ 1 30	3183·6	+ 84	1188
Umberto	− 1 30	3444·6	− 90	1014
Johnston	− 5 30	556·7	− 54	1050
Wollaston	− 4 0	3634·2	− 254	850
Alexandra Station, mean ratio	..	..	..	1302
Vittorio Emanuele	− 4° 0′	2768·1	− 194	1108
Edward	− 4 0	3298·4	− 231	1301
Elena	− 5 0	1134·6	− 99	1203
Margherita	+ 6 30	214·3	+ 24	1326
Wollaston	− 6 0	4209·8	− 443	859
Margherita, North Peak ..	+ 0 30	389·0	+ 3	1305

II.—Geodetic Observations.

SUMMARY C—*continued*.

Name of the Points.	Zenith Angles Observed.	Sides.	Differences of Level.	Ratios referred to Bujongolo.
Wollaston Station, mean ratio	..	..	..	861
Semper	+ 6° 30'	1278 ·8	+ 146	1007
Iolanda	+ 1 00	6968 ·2	+ 122	983
Edward	+ 7 30	1628 ·2	+ 214	1075
Stairs	− 1 30	2969 ·9	− 78	783
Stuhlmann	− 7 00	3748 ·8	− 460	401
Sella	0	3815 ·6	0	861
Moore	− 1 00	299 ·1	− 5	856
Iolanda Station, mean ratio	..	..	..	971
Bottego	− 3° 0'	961 ·0	− 50	921
Camp IV, ratio determined with the barometer	..	..	..	710
Grauer Pass	0°	0	..	710
Umberto Station, mean ratio	..	..	..	1005
Kraepelin..	− 0° 8' 30''	819 ·3	− 2	1003
Edward Station, mean ratio..	..	..	..	1071
Vittorio Emanuele .. ,. ..	+ 0° 30'	3894 ·9	+ 34	1105
Moore	− 7 30	1657 ·9	− 218	853
Weismann	− 3 30	3386 ·4	− 208	863

Appendix B.

SUMMARY D.

EPITOME OF THE HEIGHTS OF THE POINTS REFERRED TO BUJONGOLO AND TO SEA-LEVEL, DETERMINED BY DISTANCES AND THE BAROMETER. RATIO OF BUJONGOLO ABOVE THE SEA, 3,798 METRES (12,461 FEET).

| Points. | Ratios obtained from the Stations. | | | | | | | | | | | | | Mean. | Ratio obtained with the Barometer. | Ratios adopted and referred to. | | |
	A.	B.	Cagni.	Margherita.	Alexandra.	Wollaston.	Vittorio Emanuele.	Stairs.	Edward.	Elena.	Iolanda.	Umberto.	Camp IV.			Bujongolo. Metres.	At sea-level. Metres.	At sea-level. Feet.
Cagni	719	724	—											721	—	721	4519	14826
Edward	1072	1070	1072											1071	1078	1075	4873	15987
Vittorio Emanuele			1104						1105					1104	1102	1108	4901	16079
Margherita			1328		1326									1327	1327	1327	5125	16814
Elena			1204		1203									1203	1193	1197	4995	16388
Iolanda			957			983								970	972	971	4769	15646
Umberto			997				1014							1005	1029	1017	4815	15797
Wollaston			866						857					861	—	861	4659	15286
Sella			864		859	861		859	853					861	—	861	4659	15286
Moore			859			856								856	—	856	4654	15269
Stairs			799			783								792	—	792	4590	15059
E			1144											1144	—	1144	4942	16214
Semper		998				1007								1002	1060	1031	4829	15843
A		10												10	—	10	3808	12493
Alexandra				1302			1298		1302	1303				1302	1312	1307	5105	16749
Margherita, N. Pk.					1305									1305	—	1305	5103	16743
Savoia							1188							1188	1176	1182	4980	16339
Kraepelin												1003		1003	—	1003	4801	15752
Weismann									863					863	—	863	4661	15292
Johnston							1050							1050	—	1050	4848	15906
Stuhlmann						401								401	390	395	4193	13757
Grauer Pass													710	710	724	717	4515	14813
Bottego											921			921	—	921	4719	15482
B														—	104	104	3902	12802

III.—REPORT ON METEOROLOGICAL AND ALTIMETRIC OBSERVATIONS MADE BY H.R.H. THE DUKE OF THE ABRUZZI'S EXPEDITION TO RUWENZORI (1906).

By Prof. DOMENICO OMODEI.

The meteorological observations were made during the whole journey, from 16th May to 12th August, for the twofold purpose of first giving an idea of the climate of these regions, at least for the short time of stay, and then of determining, at least approximately, the altitudes of the various places, and especially of the more important points of the Ruwenzori Range, which was the chief objective of the expedition.

The instruments used in the observations consisted of three mercurial barometers * of the Fortin type, one registering barometer, three aneroids, two hypsometric thermometers, three thermometers and one psychrometer.

Before starting, these instruments were carefully compared with the normal instruments, and to all the data, which are recorded farther on, were applied the terms of correction thus established.

During the first part of the journey from Entebbe † to Fort Portal, from 16th to 28th May, the observations for pressure, temperature and humidity, as well as those relating to the state of the weather, were taken every day at noon, while other less complete observations were made at 15 and 21 o'clock. The summary of the observations is contained in the accompanying Table I. The maximum and minimum temperatures have reference to the whole period of the stay of the expedition in a given station, that is, generally from about 10 or 11 in the morning till the first antemeridian hours of the next day.

---

* Two graduated from 720 to 240 mm., and one from 480 to 290 mm. for the uplands.

† Here there is an observatory, the altitude of which is known, hence it was taken as the basal station for the determination of the altitudes of places between Entebbe and Fort Portal.

Appendix B.

At the Entebbe Observatory the meteorological observations were regularly carried out three times in the day, at 7, 14 and 21 of mean local time, which differs 2 hours, 8 minutes and 45 seconds from mean Greenwich time, whereas those of barometric pressure during the journey could be made only at noon of local time.

Hence in the absence of synchronous corresponding data of Entebbe, for the pressure the mean of the hours 7, 14 and 21 has been assumed, and from this mean the pressure at noon may, generally speaking, be taken to differ but slightly. For the temperature and the tension of aqueous vapour that of the nearest hour, that is 14, has been assumed. In Table II have been brought together the data of the observations made at Entebbe * between 16th and 28th May. From the data supplied by the two Tables I and II were calculated the altitudes of the various stations relatively to Entebbe by means of the formula : †

$$Z = 18400 \ (1{,}00157 + 0{,}00367 \ \theta) \left(\cfrac{1}{1 - 0{,}378 \ \dfrac{\phi}{\eta}} \right)$$

$$(1 + 0{,}00259 \ \cos 2 \ \lambda) \left(1 + \frac{Z + 2 \ z}{6371104} \right) \log \frac{H_0}{H}$$

where Z expresses the difference of level between the two stations.

H_0 the pressure reduced to 0^0 in the lower station.

H the pressure reduced to 0^0 in the upper station.

$\theta = \dfrac{t_0 + t,}{2}$ the mean between the temperature t_0 of the air in the lower station and that of t in the upper station.

$\phi = \dfrac{f_0 + f}{2}$ the mean between the vapour tension f_0 in the lower station and f that of the upper station.

$\eta = \dfrac{H_0 + H}{2}$

λ the latitude,

z the latitude of the lower station above sea-level.

The headings H_0, t_0, f_0, H, t, f, which figure above the columns in the following tables, refer to the use of the formula for the calculation of which use has been made of the "Tables Météorologiques internationales" (Paris, 1890).

---

* According to the certificate of the " National Physical Laboratory," the barometer of the Entebbe Observatory has a correction of − 0·001 inch.

† This formula of Rühlmann has been adopted without more ado as the most general and complete, since a discussion on the choice of altimetric formulas, which should take account of the recent results on the law of variation of the meteorological elements in the open air and on the slopes of the mountains, would not be in accord with the few available data, data which, moreover, cannot always be obtained under the best conditions.

III.—Meteorological, etc., Observations.

The following are the results obtained* :

Altitude of Bweya	... relatively to Entebbe m.	71	about	232·8	ft.
„ Mitiana	... „ „ „ „	60	„	196·18	„
„ Bujongo	... „ „ „ „	59	„	177·16	„
„ Bimbye	... „ „ „ „	14	„	46·0	„
„ Kijemula	... „ „ „ „	84	„	275·60	„
„ Madudu	... „ „ „ „	151	„	495·41	„
„ Muduma	... „ „ „ „	113	„	370·74	„
„ Lwamutukuza	„ „ „ „	136	„	446·20	„
„ Kichiomi	... „ „ „ „	188	„	556·81	„
„ Misongo	... „ „ „ „	125	„	410·11	„
„ Kaibo	... „ „ „ „	199	„	652·86	„
„ Butiti	... „ „ „ „	298	„	977·73	„

As it is almost needless to state, these altitudes have very uncertain value, either because obtained by isolated observations made at different hours, or else because to very slightly different levels correspond considerable horizontal distances, as between Entebbe and Fort Portal (about 225 kilometres = nearly 140 miles).

But for this last inconvenience, greater probabilities of accuracy would be presented by the determination of the altitude of Fort Portal. Here there is an observatory which has been recently founded, but the height of which relatively to Entebbe has not yet been determined geodetically, but was obtained from the observed data during the whole quarter, May, June, and July, 1906, so that the influence of the various sources of error may be regarded as considerably lessened.

In the subjoined Table III are recorded all these data, the outcome of which was that the difference of altitude between Entebbe and Fort Portal† is 355 metres (1,170 feet).

---

* The calculation has been omitted for Katende which, as shown by the pressure, stands at very nearly the same level as Entebbe.

† From the comparisons made on the 30th and 31st May, and again on the 21st and 24th July between the two Fortin barometers of the expedition and the barometer No. 2,025 (Negretti and Zambra) of the English Observatory at Fort Portal, it appeared that the former gave a mean indication of 5·17 mm. = ⅕ inch (at 0°) above that of the latter. Since this difference was verified in an equal degree with the two barometers of the expedition, which kept in perfect agreement with each other at Fort Portal, as they had before the journey, it was thought reasonable to apply to all the pressures yielded by the barometer of the Fort Portal Observatory, the constant correction 5·17 mm. To the end of November, 1907, the correction of the Kew Observatory had not yet been obtained for the Fort Portal barometer.

Appendix B.

A result quite conformable (the difference is less than half a metre, or 18 or 19 inches) is reached by assuming for the calculation the means of the data of Entebbe and Fort Portal for seven months of the year 1905 (*see* Table IV).

It was impossible to include the whole year, because no observations were made at Fort Portal from May to September, 1905. Hence, the Entebbe Observatory being 3,863 English feet, or 1,177 metres above sea-level, it follows that Fort Portal stands at about 1,532 metres (5,025 feet) above the sea.

After a stay of two days at Fort Portal, the expedition started on 1st June from Fort Portal for Ruwenzori.

In the appended Table V are given the data of the observations made at the various encampments, as in Table VI the corresponding data of Fort Portal, where, at the request of H.R.H., Mr. John de Souza, Director of the Observatory, besides the ordinary observations for the hours 7, 14, and 21, made one also at noon for the whole time that elapsed between the departure of the expedition from, and its return to, Fort Portal.

From the data of Tables V and VI have been calculated the following altitudes :

Altitude of	Duwona	relatively to Fort Portal about m.						54 =	177·0 ft.
,,	Kasongo	,,	,,	,,	,,	,,	,,	136 =	446·2 ,,
,,	Ibanda	,,	,,	,,	,,	,,	,,	458 =	518·3 ,,
,,	Bihunga	,,	,,	,,	,,	,,	,,	388 = 1,273·0 ,,
,,	Nakitawa	,,	,,	,,	,,	,,	,,	1,120 = 3,674·6 ,,
,,	Kichuchu	,,	,,	,,	,,	,,	,,	1,465 = 5,788·2 ,,
,,	Buamba	,,	,,	,,	,,	,,	,,	1,986 = 6,515·8 ,,

On 8th June the expedition reached Bujongolo, a place which is comprised within the Ruwenzori uplands, and as this formed the basal station and point of reference for all the measurements to be subsequently taken during the exploration, the observations were here made regularly from 16th June to 12th July, under conditions far more favourable than those that would be secured while *en route*. The instruments were suspended from a vertical table supported by two posts, which were firmly planted in the ground at a height of about five feet above the surface, and protected from the effects of insulation and of the rain by a large awning extended above at a distance of about a foot.

At Table VIII are given the results of these observations, which are recorded in full, not only because they have been used for the calculation of the

III.—Meteorological, etc., Observations.

altitudes, but also because they serve to give an idea of the climate of* that interesting locality.

Then at Table VII are brought together the data of Fort Portal for the same period of time, bearing in mind that to the pressures reduced to 0° has been applied the already-mentioned constant term of correction + 5·17 mm. Then from the mean data of Tables VII and VIII was calculated the difference of level between Bujongolo and Fort Portal. The first calculation was made with the data of the synchronous midday observations at Bujongolo and Fort Portal, and was found to be 2,276·7 *metres* (7,468·5 feet).

But when we allow for the considerable difference of level between the two observed stations, we cannot assert with certainty that the law of daily variation of pressure is identical in both places, hence the coincidence of the hour of observation does not imply identity of modifications in the atmospheric ebb and flow, so that it becomes advisable to try and take advantage of the other observed data too, besides those of midday. Therefore with the mean daily values of pressure, temperature,† and vapour tension for Fort Portal (obtained from the mean of the three observations of the hours 7, 14, and 21 for Bujongolo) are associated the mean values of the pressure and vapour tension deduced from the two observations of the hours 9 and 17. These, when account is taken of the normal movement of the daily variations of pressure and tension, should not differ greatly from the diurnal mean.

For the temperature of Bujongolo we have assumed the mean of the maximum and minimum temperature, which, in the absence of more complete data, is the one that approaches nearest to the mean daily temperature.

With the values thus obtained, and recorded at foot of Tables VII and VIII, the difference of level has been calculated between Fort Portal and Bujongolo, and is found to be 2,255·7 metres (7,376 feet).

Taking as a more approximate value the mean between this and the preceding value, we get as the height of Bujongolo above Fort Portal *2,266 metres* (7,432 feet), and adding to this value the altitude of Fort Portal above the sea, the elevation of Bujongolo above the sea is found to be *3,798 metres* (12,461 feet).

The camp being established at Bujongolo, where, as already stated, regular observations were taken three times daily from 15th June to 12th July, the excursions began to the chief places in the Ruwenzori Range. The instruments

---

* Owing to the requirements of daily life at the Bujongolo encampment, the meteorological observations were taken at the hours of 9, 12, and 17, instead of 7, 14, and 21, as at Fort Portal and Entebbe.

† In the calculation of temperature no account is taken of the maximum and minimum, because at times they disagree with the other temperatures of the day.

Appendix B.

brought with us on these excursions were :—a Fortin barometer which had for a long time been compared with another left at Bujongolo ; an aneroid likewise compared with the two Fortins ; a thermometer for taking the temperature of the air, and two hypsometric thermometers.

For the more important points, for instance, for nearly all the peaks, and always where possible, the measurements of pressure were made with the mercurial barometer, the aneroid being used only in a few special cases where it would have been very difficult to carry, or take measurements with the mercurial barometer, and also for places of secondary importance. The precaution, however, was taken to take down or record the indications of the aneroid even whenever the Fortin was used.

Not till after 12th July, when the Fortin got damaged, was the hypsometer employed. Whenever it was possible, the observations were made at the same hours as those of Bujongolo (9, 12, and 17), but occasionally this was not possible, and then we assumed as terms of comparison the data of Bujongolo made at the nearest hours, unless there were reasons for adopting the mean of two consecutive data.

In connection with this preferable choice of data for the calculation of altitudes it should be noted that for places for which the daily variation of pressure, temperature, etc., is known, that is to say, where the hour of the maxima and minima and the extent of the daily variations are ascertained, it is possible to reduce a determination made at any given hour to another determined hour. But in the present case these fundamental notions are lacking, and for regions such as that under consideration, meteorological studies are too rare to enable us confidently to extend to them those laws that have been established for regions of the temperate zone.*

For Bujongolo we should no doubt have some element to establish approximately the daily movement of pressure, temperature, etc., but the same cannot be said for the other places in Ruwenzori, and especially for the peaks.

For these reasons the heights were calculated with the data, such as they were, without modifying them in any way on the ground of the hours when the observations were made.

Another matter, which, however, has no great influence, is that concerned with the humidity, or rather the tension of the aqueous vapour which is always

---

* Let one example suffice to show what caution is necessary in this respect. At Bujongolo the mean pressure at 9 o'clock is 488·87 mm. (see Table VIII, 6) ; at 12, 488·67, and at 17, 488·08, so that the maximum of the morning is reached before midday, and at this hour the barometer is already falling. On the other hand on the Säntis Peak (2,467 metres), the maximum of the morning is delayed till toward 14 o'clock, and on Mt. Blanc (4,811 metres), till towards 15 (Angot, météorologie).

III.—Meteorological, etc., Observations.

found to a greater or less extent in the atmosphere. In the formula which serves for the calculation of the altitudes (*see* p. 362) there occurs the factor $\dfrac{1}{1-0{,}378\,\dfrac{\phi}{\eta}}$ where ϕ is the mean vapour tension at the two stations, and η the mean of the pressures, and this factor has reference to the influence exercised by the presence of the aqueous vapour on the readings of barometric altitudes. For Bujongolo the vapour tension is known, but not for the observed places on Ruwenzori, as here no psychrometric observations were made.

The neglect of the factor relating to the humidity might be a cause of error, to eliminate which, at least partly, a mean humidity of about 60 has been admitted for the stratum of the air comprised between Bujongolo and the observed station.*

This humidity of 60 is certainly less than the true mean, since at Bujongolo the humidity is always very high (mean 89), and there is reason to believe that it is always considerable in the other places too, where cloudy, foggy and rainy weather prevail.

On the Tables IX, X, XI and XII are recorded the altitudes of the various other places on Ruwenzori, calculated with the previously indicated *normæ*.

Regarding the results obtained, it may be noticed that the determinations made with the mercurial barometer were found to agree sufficiently well with each other whenever it was possible to make more than one determination for any given place, and they agree also with the surveys made with geodetic methods.

Owing to the irregular behaviour of the aneroid barometers, the measurements taken with these instruments present a far less degree of approximation.

---

* To show the possible influence of such a correction, reference may be made to the special case of the Margherita Peak. Here the pressure at 11 o'clock on 18th June was 414·0 mm. and the temperature − 3°·3 Celsius (26°·6 F.), whereas at Bujongolo, at 12 o'clock on the same day, the pressure was 487·9 mm. and the temperature 5°·1 Celsius (41° F.), hence the mean pressure was about 451 mm. and the mean temperature 0°·9 Celsius (33° F.). Had the air been saturated at this temperature the vapour tension would have been 4·87 mm. Admitting a humidity of 60 the tension falls to 2·92 mm., with which datum, and with the mean pressure of 451, we get the cologarithm of the term of correction for the humidity, namely :

$$\text{colog.}\ \frac{1}{1-0{\cdot}378\,\dfrac{\phi}{\eta}}=0{\cdot}00103$$

Without taking account of the humidity, the height of Margherita Peak above Bujongolo was found to be 1,324 metres; with this added it becomes 1,327 metres, that is to say, we have a rise of about 0·22 per cent. Admitting a humidity of 80 the height would become 1,328·5 metres, with a rise of 0·33 per cent.

Appendix B.

In fact, in spite of every care taken to make continual comparisons with the Fortin barometer, the term of correction did not keep constant, not only from day to day, but even during the same day whenever the instrument got shaken or was exposed to sudden changes of altitude.

Hence, to avoid errors that might even be serious, a cautious and limited use has been made of the data obtained with the aneroid. Thus, the simultaneous indications of the Fortin and the aneroid being noted, and the altitude obtained from the former being taken as correct, the indications of the latter have served to establish differences of level, which were inconsiderable relatively to the point of comparison. When this process was completed the comparison was renewed, so that the data first obtained were brought under control.

In general the reported data result from the mean of two or more determinations, and may consequently be regarded as sufficiently accurate.

As to the altitudes of the places passed by the expedition on the return journey, that is, from and after the 14th July, these were all obtained by means of comparisons with the data observed simultaneously at Ibanda.* Then, for the sake of uniformity, they were reduced, like the previous ones, to the common level of Bujongolo.

Amongst the determinations made relatively to Ibanda was that of Iolanda Peak, the altitude of which will consequently not be so near the truth as that of the other peaks. And, besides the inconvenience of Ibanda lying still lower than Fort Portal, there was also the trouble caused by the breaking of one of the mercurial barometers, instead of which we had to use the hypsometer, which yields a less degree of approximation in the readings.

NOTE.

In the western districts of the Uganda Protectorate the geodetic survey has not yet been carried out, but the far-seeing British Government is taking it in hand, and no doubt it will soon be an accomplished fact. Hence it might seem reasonable to refer the various altitudes of the Ruwenzori group, not to Bujongolo, but to Fort North Portal, the exact height of which above sea-level will soon be known. In fact, this very critical point is being seen to by Messrs. H. Y. Tegart and H. E. Maddox, who have recently published some

---

* For the altitude of Ibanda relatively to Fort Portal – 148 was assumed, this being the mean of the observations made on going and returning.

III.—Meteorological, etc., Observations.

valuable studies on Ruwenzori. But if the data have nevertheless been referred to Bujongolo, it was because they offered greater guarantees of accuracy, and this for the following reasons :—

First of all, the difference of altitude between Bujongolo and Fort Portal may be regarded as near enough, having been obtained from a considerable number of observations. On this account we may consider as partially compensated those sources of error which are due to atmospheric disturbances, such as may have a great influence on isolated measurements, especially when dealing with very remote stations. Such would precisely have been the case if the data observed on Ruwenzori had been directly compared with those corresponding hour for hour with Fort Portal. But by making the comparisons with the data obtained at Bujongolo, a much nearer place, one may fairly assume a greater uniformity of atmospheric conditions.

Then there is another fact which shows the greater convenience of the course adopted. It is seen in the following example to which many others might be added.

On 7th July, at 12 o'clock, on Edward Peak, the pressure (reduced to $0°$) was 428·5 mm., and the temperature $1°·6$ Celsius ($34°·4$ F.), the corresponding readings being at Bujongolo 489·16 mm. and $3°·9$ Celsius ($39°$ F.), and at Fort Portal 638·69 mm., and $23°·3$ Celsius ($74°$ F.).

Calculating from these data the difference of level between Edward Peak and Fort Portal, and then separately between Edward Peak and Bujongolo, and between Bujongolo and Fort Portal, we get :—*

Difference of level between Edward Peak and Fort Portal ... 3,355 metres
Difference of level between Edward Peak and
    Bujongolo   ...   . 1,074 metres
Difference of level between Bujongolo and Fort
    Portal...   2,249   ,,

Total difference of level between Edward Peak
    and Fort Portal   ...   . 3,323 metres (10,900 feet)

Therefore, with the direct calculation, and omitting Bujongolo, we have a difference of over 32 metres (105 feet).

Such a difference arises from the fact that whereas the law of Laplace is based on the hypothesis of a static equilibrium of the atmosphere, and of a temperature and humidity which decrease regularly with the altitude, this does not take place in the present instance.

---

* In this estimate no account is taken of the tension of the aqueous vapour.

Appendix B.

In fact, the temperature being 23°·9 Celsius (75° F.) at Fort Portal, and 1°·6 Celsius (34°·4 F.) at Edward Peak, a difference of 22°·3 Celsius (72°·6 F.), if the decrease occurred proportionately to the altitude the temperature of Bujongolo should be about 7°·2 Celsius (45° F.), whereas it is only 3°·9 Celsius (39°·7 F.). This, therefore, means that the column of air has a lower temperature* than is assumed by the theory, so that to an equal difference of pressure corresponds a less difference of altitude.

Lastly, in connection with determinations of this nature, it is not to be forgotten that results now well established are :—

1. That heights calculated by means of thermo-barometric observations are generally found to be greater with measurements made by day compared with those made by night. They present a considerable daily range with the maximum value shortly before the maximum daily temperature, and the minimum one or two hours before sunrise. The extent of range is influenced by the season, the local conditions, and the state of the sky.

2. That the altitudes calculated with the mean daily or monthly values of the observations are found to be too low in winter and too high in summer ; with the annual mean they differ little from the actual heights.

---

* Caused probably by the great masses of ice on Ruwenzori.

III.—Meteorological, etc., Observations.

TABLE I.

OBSERVATIONS MADE AT THE STATIONS COMPRISED BETWEEN ENTEBBE AND FORT PORTAL.

Number.	Stations.	Date. Month.	Date. Day.	Hour.	Pressure reduced to 0°. H.	Temperature. Noon.	Temperature. Maximum.	Temperature. Minimum.	Vapour Tension. f.	Wind.	Sky.	Atmosphere.	Remarks.
1	Katende	May	16	12	664·88	27·5	29·5	19·0	17·18	calm	clear	bright	Rain at night without thunder.
2	Bweya	,,	17	12	659·25	29·2	—	—	15·65	calm	clear	bright	
3	Mitiana	,,	18	12	660·60	28·7	—	16·0	14·68	calm	overcast	dull	Fog in morning; lightning in N.W.
4	Bujongolo	,,	19	12	661·16	30·7	31·0	16·0	15·35	calm	clear	—	Overcast on leaving Bujongolo; then clear.
5	Bimbye	,,	20	12	664·04	27·7	28·0	12·0	16·58	S.E. light	half overcast	bright	Lightning at night, 4 and 1 quadrant. Cumuli.
6	Kijemula	,,	21	12	659·06	25·7	26·0	14·0	15·69	S.E. high	cloudy	dull	Rain at night; fog morning on leaving Kijemula.
7	Madudu	,,	22	12	653·38	28·2	29·5	15·0	20·39	S.E. light	half overcast	dull	Light rain morning; shifting clouds. Cumuli, cirri, strata, fog.
8	Muduma	,,	23	12	655·98	23·2	29·0	12·0	15·79	calm	half overcast	bright	Heavy dew.
9	Lwamutukuza	,,	24	12	654·95	28·7	31·0	16·0	21·20	calm	half overcast	bright	Light dew. Shifting from clear to overcast.
10	Kichiomi	,,	25	12	650·59	24·2	26·0	15·5	17·41	S.W. high	half overcast	bright	At 15 storms S.E. and S.W., at 18½ wind E., and storm, with gusts, rain, lightning, thunder.
11	Mujongo	,,	26	12	655·32	26·7	29·0	17·0	22·50	S.W. light	clear	dull	Rain at 4 on leaving, shifting from clear to overcast.
12	Kaibo	,,	27	12	649·81	23·3	26·5	15·0	15·55	S.W. light	half overcast	dull	At 5 cloudy, then clear. Ruwenzori sighted for the first time.
13	Butiti	,,	28	12	643·13	22·0	22·5	15·0	17·19	—	—	—	

371

Appendix B.

TABLE II.—RECORD OF OBSERVATIONS AT ENTEBBE FROM 16TH TO 28TH MAY, 1906.

Date.	Pressure reduced to 0°.				Temperature.			Vapour Tension.		
	Hour 7.	Hour 14.	Hour 21.	Mean.	Hour 7.	Hour 14.	Hour 21.	Hour 7.	Hour 14.	Hour 21.
16th May ...	665·36	664·47	664·54	664·79	21·9	25·6	21·5	16·40	16·73	15·81
17th ,, ...	665·24	664·20	664·04	664·49	22·2	25·4	22·2	15·04	16·14	16·04
18th ,, ...	665·54	664·65	664·88	665·02	22·5	25·8	20·7	15·35	15·72	15·30
19th ,, ...	666·62	664·66	665·17	665·48	21·6	26·5	21·1	15·07	18·54	16·06
20th ,, ...	665·52	664·36	665·25	665·04	21·4	26·1	21·8	15·53	17·32	16·29
21st ,, ...	665·88	665·00	665·25	665·37	22·3	24·5	21·2	16·32	16·76	16·65
22nd ,, ...	665·74	663·30	664·62	664·55	22·3	25·4	21·4	17·01	17·03	17·05
23rd ,, ...	664·67	664·58	664·05	664·43	17·7	23·9	21·9	14·46	16·20	16·05
24th ,, ...	666·13	664·45	664·40	664·99	18·2	24·9	21·6	14·45	16·80	16·58
25th ,, ...	665·30	663·95	664·61	664·62	18·3	22·8	21·2	14·86	17·51	17·00
26th ,, ...	664·81	664·96	664·00	664·61	18·7	22·2	21·9	15·41	18·47	15·72
27th ,, ...	665·78	664·23	664·00	664·66	18·4	24·0	22·1	15·43	18·24	16·27
28th ,, ...	665·29	665·77	664·61	665·22	17·2	21·5	21·4	13·85	18·37	17·05

NOTE.—According to the Certificate of the London "National Physical Laboratory," the barometer of the Entebbe Observatory has a correction of − 0·001, and of this account has been taken in the data reported in this table.

TABLE III.—ENTEBBE.

MONTHS.	Mean Monthly Pressure reduced to 0° at the hours			Mean Monthly Temperature at the hours			Mean Monthly Vapour Tension at the hours		
	7	14	21	7	14	21	7	14	21
	mm.	mm.	mm.				mm.	mm.	mm.
May	665·48	664·29	664·57	20°·1	24°·2	21°·4	15·51	16·88	16·37
June...	665·63	664·55	664·85	18·5	22·9	20·7	14·59	16·30	15·80
July	665·61	665·04	665·18	17·7	23·7	20·7	12·65	15·35	14·98
Means for the quarter	665·57	664·62	664·87	18·8	23·6	20·9	14·25	16·18	15·72
Means for the hours	665·02			21°·1			15·39		

FORT PORTAL.

MONTHS.	Mean Monthly Pressure reduced to 0° at the hours			Mean Monthly Temperature at the hours			Mean Monthly Vapour Tension at the hours		
	7	14	21	7	14	21	7	14	21
	mm.	mm.	mm.				mm.	mm.	mm.
May	638·08	637·20	638·05	16°·4	23°·6	16°·7	11·99	13·04	12·67
June...	638·85	637·62	638·58	14·7	22·4	16·4	10·67	12·04	12·27
July	639·47	638·24	639·28	13·3	23·5	15·7	9·97	11·99	11·43
Means for the quarter	638·80	637·69	638·64	14·8	23·1	16·3	10·87	12·32	12·12
Means for the hours	638·37			18°·1			11·7		

TABLE IV.   ENTEBBE.   FORT PORTAL.

Months.	Hour.	Pressure in inches.	Associated therm. F.	Psychrometer. Dry therm. F.	Psychrometer. Wet therm. F.	Pressure in inches.	Associated therm. F.	Psychrometer. Dry therm. F.	Psychrometer. Wet therm. F.
			°	°	°		°	°	°
January ..	7	26 ·281	66 ·8	66 ·3	64 ·8	24 ·969	59 ·3	57 ·6	56 ·1
,, ..	14	·245	77 ·9	76 ·0	70 ·4	24 ·950	73 ·6	74 ·3	65 ·6
,, ..	21	·257	69 ·4	69 ·1	66 ·6	24 ·956	62 ·7	61 ·7	59 ·5
February..	7	·276	67 ·7	66 ·8	64 ·3	24 ·964	59 ·6	58 ·0	55 ·1
,, ..	14	·236	81 ·3	79 ·1	71 ·3	24 ·950	76 ·4	76 ·1	64 ·1
,, ..	21	·229	70 ·5	70 ·0	67 ·3	24 ·954	61 ·7	60 ·7	57 ·7
March ..	7	·264	68 ·6	67 ·8	65 ·1	24 ·973	62 ·1	61 ·1	53 ·8
,, ..	14	·217	76 ·8	74 ·9	68 ·9	24 ·943	75 ·0	73 ·3	65 ·2
,, ..	21	·245	70 ·6	69 ·9	66 ·9	24 ·958	63 ·0	62 ·5	60 ·3
April ..	7	·299	69 ·4	68 ·5	66 ·0	24 ·995	62 ·2	60 ·6	58 ·8
,, ..	14	·244	77 ·1	74 ·7	69 ·5	24 ·974	78 ·3	75 ·4	66 ·3
,, ..	21	·244	69 ·8	69 ·9	66 ·9	24 ·979	63 ·1	62 ·6	60 ·1
October .	7	·257	66 ·3	65 ·3	63 ·9	25 ·098	65 ·3	64 ·1	60 ·1
,, ..	14	·221	79 ·0	76 ·1	69 ·9	25 ·075	72 ·3	69 ·9	62 ·7
,, ..	21	·241	69 ·6	69 ·0	66 ·4	24 ·863	63 ·5	65 ·2	60 ·0
November	7	·272	65 ·7	65 ·1	63 ·9	25 ·067	65 ·9	61 ·9	59 ·5
,, ..	14	·229	77 ·8	75 ·5	69 ·4	25 ·048	72 ·9	71 ·8	65 ·1
,, ..	21	·247	69 ·5	68 ·9	66 ·6	24 ·904	63 ·0	65 ·7	59 ·0
December	7	·262	66 ·4	65 ·4	64 ·2	25 ·025	64 ·1	59 ·7	58 ·4
,, ..	14	·227	78 ·1	75 ·0	69 ·3	25 ·098	75 ·0	71 ·9	64 ·9
,, ..	21	·246	69 ·8	69 ·3	66 ·4	25 ·885	64 ·2	68 ·5	62 ·0
Means ..		26 ·648	71 ·8	70 ·6	67 ·1	24 ·982	66 ·8	65 ·8	60 ·6

Correct pressure reduced to 0° and in mm... $H_0 = 664 ·28$ $H = 637 ·70$

Temperature in centigrades   .... $t_0 = 21 ·4$ $t = 18 ·8$

Vapour tension in mm.   .... $f_0 = 15 ·7$ $f = 11 ·7$

III.—Meteorological, etc., Observations.

TABLE V.
OBSERVATIONS MADE AT THE STATIONS COMPRISED BETWEEN FORT PORTAL AND BUJONGOLO.

Number.	Stations.	Month.	Day.	Hour.	H. Pressure reduced to 0°.	t.	Maximum.	Minimum.	Vapour Tension.	Wind.	Sky.	Atmosphere.	Remarks.
1	Duwona ..	June	1	12	633·92	23·7	28·0	13·5	20·02	calm	clear	bright	Wind S.E. light; morning calm, then fresh later, and clear.
2	Kasongo ..	„	2	12	647·45	26·7	28·0	16·5	20·91	calm	clear	dull	Sky clear and overcast alternately.
3	Ibanda ..	„	3	12	650·00	25·2	31·0	14·0	19·99	calm	clear	bright	Weather very fine morning, some clouds afternoon.
4	Bihunga ..	„	4	12	611·20	20·7	23·5	14·0	15·12	S. light	clear	bright	Some clouds and darkness below in the afternoon.
5	Nakitawa ..	„	5	12	561·25	13·7	19·5	9·0	9·55	calm	overcast	dull	Light clouds at 6; clear at sunrise.
6	Kichuchu ..	„	6	12	538·93	12·2	19·5	13·5	10·46	calm	rain	fog	Light rain at noon.
7	Buamba ..	„	7	12	506·66	11·7	—	—	5·69	calm	cloudy	bright	Overcast and foggy at intervals.

TABLE VI.—CORRESPONDING DATES OF OBSERVATION AT FORT PORTAL.

	Pressure reduced to 0°. H_{o}.	Temperature. t_{o}.	Vapour Tension. f_{o}.
June 1, hour 12 : :	637·81	25·0	11·55
„ 2 „ „ : :	637·62	25·0	10·76
„ 3 „ „ : :	638·47	24·4	10·20
„ 4 „ „ : :	638·90	23·9	11·44
„ 5 „ „ : :	638·62	26·1	10·87
„ 6 „ „ : :	639·16	21·7	12·78
„ 7 „ „ : :	638·39	23·3	10·87

Appendix B.

TABLE VII.—GENERAL VIEW OF THE OBSERVATIONS MADE AT BUJONGOLO FROM 15TH JUNE TO 12TH JULY, 1906.

Month	Day	Pressure reduced to 0°. Hour 9	Hour 12	Hour 17	Temperature. Maximum	Minimum	Hour 9	Hour 12	Hour 17	Vapour Tension. Hour 9	Hour 12	Hour 17	Relative Humidity. Hour 9	Hour 12	Hour 17	Notes on the Weather.
June	15	489·64	489·36	489·84	7·0	0·8	4·4	6·9	4·9	5·29	6·70	5·29	84	92	84	Cloudy and light wind.
	16	489·82	489·28	489·04	8·0	1·3	5·9	6·1	4·9	5·00	5·61	5·50	74	79	84	Half overcast.
	17	489·49	489·08	488·26	14·0	1·3	7·9	8·9	5·9	6·35	5·75	5·72	79	67	84	Nearly clear.
	18	489·44	487·90	487·62	12·0	1·5	2·9	5·1	5·0	5·20	6·41	5·65	91	97	86	Cloudy.
	19	489·14	488·06	487·62	14·1	1·3	7·1	9·1	5·9	5·28	7·42	5·94	70	86	85	Cloudy.
	20	489·06	487·94	487·02	11·0	0·3	6·1	5·1	5·9	5·88	5·50	6·47	83	85	98	Nearly clear, then overcast.
	21	487·58	487·02	486·22	12·0	1·3	5·9	5·1	4·9	5·65	5·50	6·08	86	84	100	Overcast morning, then rain and fog.
	22	487·22	487·02	486·14	12·0	1·5	5·1	5·1	5·9	6·05	6·53	6·58	88	86	100	Overcast, dense fog evening.
	23	487·02	487·22	486·38	11·5	2·3	5·1	5·1	5·0	6·47	6·53	6·53	98	100	100	Overcast.
	24	488·82	489·44	488·82	10·8	2·3	5·7	5·4	4·6	6·54	6·32	6·36	98	97	97	Morning overcast and foggy, then light rain.
	25	489·69	489·64	488·76	7·3	2·3	5·7	5·4	5·2	6·66	6·44	6·41	95	95	95	Fog and light rain.
	26	488·76	489·10	488·02	11·5	1·1	4·2	4·7	4·7	5·92	6·01	5·92	91	92	97	Fog.
	27	488·59	489·47	487·70	7·5	1·3	5·9	6·7	6·1	6·17	6·70	6·69	82	95	97	Fog.
	28	490·14	489·62	488·98	10·2	2·3	6·7	4·9	5·1	6·05	5·92	6·32	81	100	89	Sunshine till 9, then dense fog, hail at 11.
	29	489·25	488·60	488·38	6·0	2·1	5·5	4·9	5·5	5·38	6·31	5·81	92	95	92	Sunshine till 8.30, then fog and light rain.
	30	488·80	488·10	487·98	—	2·1	6·1	4·9	4·1	6·47	6·01	5·59	97	98	93	Dense fog and light rain.
July	1	489·08	488·28	487·82	5·2	2·8	3·9	4·5	4·6	5·98	6·00	5·80	83	91	97	,, ,,
	2	489·67	489·08	488·64	6·5	1·5	5·1	6·2	4·4	5·44	6·41	5·78	100	97	95	Fog and rain.
	3	488·80	488·96	489·27	6·0	1·5	4·9	5·1	3·6	6·45	6·32	—	95	95	—	,, ,,
	4	489·32	489·10	488·42	5·4	2·3	5·9	4·9	3·7	6·44	6·01	5·57	92	95	87	Sunshine from 8 to 9, then fog till 17.
	5	489·66	489·48	488·68	7·0	0·1	3·9	4·7	3·9	5·59	5·80	5·59	97	67	98	Fog.
	6	488·88	—	—	7·0	1·5	3·9	—	—	5·57	—	—	90	—	—	Dense fog.
	7	489·14	489·16	488·70	7·0	2·1	4·9	3·9	4·9	5·83	6·67	5·71	86	69	87	Cloudy and clear alternately.
	8	488·09	487·82	487·18	7·0	1·3	4·9	4·1	4·1	5·60	5·22	5·63	86	74	98	Half overcast.
	9	487·46	488·46	488·80	—	2·3	5·9	10·9	5·0	6·08	7·31	5·71	90	66	87	Clear, overcast towards evening.
	10	488·98	487·84	487·98	—	1·3	7·9	8·9	4·1	4·74	5·50	6·04	86	85	98	Clear, then fog and overcast.
	11	488·42	487·94	487·98	11·0	2·3	5·9	6·1	5·0	4·30	6·40	6·53	59	97	100	Clear, then overcast.
	12	488·46	488·88	488·02	—	0·3	9·1	6·1	5·1	8·45	6·88	6·17	57		91	,, ,,
Mean	—	488·87	488·61	488·08	9·0	1·6	5·6	6·1	4·8	5·88	6·23	5·75	87	88	93	

Mean pressure from 9 to 17 ...   488·47. mm.
Vapour tension from 9 to 17 ...   5·81. mm.
Mean temperature (maximum and minimum)...   5·3.

376

III.—Meteorological, etc., Observations.

Table VIII.

Number.	Place.	Corresponding Number of the Map.	Month and Day.	Hour.	Type of Barometer.	Altitude. Above Bujongolo. Each separately. m.	Mean. m.	Above the Sea.	Temperature. Centigrades.	State of the Weather.
1	Margherita Peak	..	18 June	11	Fortin	1327	1327	5,125 m. 16,813 ft.	−3·3	Wind S.E. fresh ; sunshine at intervals.
2	Alexandra Peak	..	18 ,,	12. 45	,,	1313	1311·5	5,110 m. 16,749 ft.	−2·3	Fine till 8 o'clock, then fog and light wind N.E. all the day. High temperature by day on the peaks.
			20 ,,	7.30	,,	1310			−0·3	
3	Elena Peak	..	20 ,,	12	,,	1193	1193	4,991 m. 16,388 ft.	5·7	
4	Savoia Peak	..	20 ,,	14	,,	1176	1176	4,974 m. 16,339 ft.	5·7	
5	Vittorio Emanuele Peak	..	23 ,,	9	,,	1096	1102	4,900 m. 16,079 ft.	0·2	Fine till 6.30, then overcast ; snow and electric displays on the peaks.
			23 ,,	12	,,	1108			0·3	
			5 July	..	,,	1076			0·3	Clear till 8, then fog in the valleys, and later also on the peaks. At 17, after heavy snowfall, clear.
6	Edward Peak	..	7 ,,	..	,,	1076	1078	4,876 m. 15,984 ft.	1·6	Fog till 7, then clear till noon ; later fog again.
			7 ,,	..	,,	1082			0·6	
			7 ,,	..	,,	1078			−1·4	
7	Umberto Peak	..	28 June	12	,,	1029	1029	4,827 m. 15,797 ft.	0·0	Wind in variable quarters ; clouded above, bright below, till 8 ; then fog till 17, and then clear.

377

Appendix B.

TABLE VIII—continued.

Number.	Place.	Corresponding Number of the Map.	Date. Month and Day.	Date. Hour.	Type of Barometer.	Altitude. Above Bujongolo. Each separately. m.	Altitude. Above Bujongolo. Mean. m.	Altitude. Above the Sea.	Temperature. Centigrades.	State of the Weather.
8	Semper Peak ·· ··	5	10 June	8	Aneroid	1060	1060	4,858 m. 15,843 ft.	0·7	
9	Iolanda Peak* ·· ··	··	16 July	9	Hypsometer	972	972	4,770 m. 15,646 ft.	3·6	
10	Stairs Peak ·· ··	32	8 ,,							
11	Grauer Camp ·· ··	1	9 ,,	17.45	Fortin	234	234	4,032 m. 13,229 ft.	3·6	
12	Lower limits of the Moore Glacier	2 (a)	9 June	9	Aneroid	445	445	4,243 m. 13,917 ft.	4·2	
13	Camp I ·· ··	3	9 ,,	11 and 19	,,	551	551	4,349 m. 14,118 ft.	2·7	
14	Grauer Pass ·· ··	4	10 ,,	··	,,	784	784	4,582 m. 15,029 ft.	0·7	
15	Cleared Mobuku Valley	6 (b)	10 ,,	··	,,	132	132	3,930 m. 12,894 ft.		
16	Freshfield Pass ··	7	15 ,,	12	Fortin	528	528	4,326 m. 14,193 ft.	3·2	Clear till 8, then fog. Wind E.-S.E. fresh; on Pass 1 clear after 13 o'clock.
17	Camp II ·· ··	8	15 ,,	17	,,	247	247	4,045 m. 13,271 ft.	3·2	
18	Plateau above the lakes	9 (c)	16 ,,	10.30	Aneroid	273	273	4,071 m. 13,353 ft.	6·2	Overcast. Wind, 2nd quarter, light. After midday clear, with wind in 1st quarter.
19	Camp III ·· ··	11	16 ,,	17	Fortin	421	421	4,219 m. 13,842 ft.	2·7	
20	Camp IV ·· ··	12	17 ,,	12	Fortin	718	718	4,516 m. 14,817 ft.	3·7	Wind 3rd quarter; fog and sunshine at intervals.

III.—Meteorological, etc., Observations.

No.	Station	Ref.	Date	Month	Time	Instrument			Altitude	*	Remarks
21	Stanley Glacier, near Alexandra Peak	13 (d)	20	,,	9.10	Aneroid	1007	1007	4,805 m. / 15,765 ft.	..	Fog and light N.E. wind.
22	Clearing on mountain at Lake Bujuku	14	22	,,	11	Fortin	135	135	3,933 m. / 12,904 ft.	6·2	Foggy and clear at intervals.
23	Névé below Speke Glacier	15 (e)	22	,,	12	Aneroid	319	319	4,117 m. / 13,504 ft.	8·7	
24	Camp V	16	26	,,	17	Fortin	677	677	4,475 m. / 14,682 ft.	3·0	Fine till 6.30, then the peaks clouded.
25	Lower limit of Speke Glacier	17 (f)	29 / 27	,, / ,,	16 / 8	,, / Aneroid	677 / 646	646	4,444 m. / 14,576 ft.	4·0	Fog nearly the whole day.
26	Spur of N.W. crest of Mt. Speke	18 (g)	27	,,	9	,,	696	696	4,494 m. / 14,740 ft.	..	Foggy and clear at intervals. Wind constantly variable.
27	Belvedere	19 (h)	27	,,	11	,,	671	671	4,469 m. / 14,658 ft.	..	
28	Camp VI	20	27 / 28	,, / ,,	17 / 8.20	Fortin / ,,	434 / 458	446	4,244 m. / 13,924 ft.	4	
29	Cavalli Pass	21	28	,,	9	Aneroid	512	512	4,310 m. / 14,141 ft.	4	Cloudy above and clear below till 8, then fog, and afterwards clear.
30	Camp VII	22	28	,,	17	Fortin	667	667	4,465 m. / 14,649 ft.	1	
31	Lake below the Speke Glacier slope	23	29	,,	13.30	Aneroid	480	480	4,278 m. / 14,032 ft.	..	Fog nearly the whole day.
32	Stuhlmann Pass	24	30	,,	9	Fortin	390	390	4,188 m. / 13,757 ft.	4	Fog till 7, then overcast, rain and snow.
33	Scott Elliot Pass	25	30	,,	13.30	,,	549	549	4,317 m. / 14,262 ft.	4	
34	Lower limit of Valley B, Névé	26 (m)	1 July		..	Aneroid	303	303	4,101 m. / 13,484 ft.	..	
35	Camp VIII	30	5 / 7	,, / ,,	20.30 / 18	Fortin / ,,	514 / 534	524	4,322 m. / 14,180 ft.	0·6 / 0·6	Clear till 8; then fog. At 17, after a heavy snowfall cleared up on the peaks, fog.
36	Altitude N. Freshfield Pass	31 (p)	6	,,	17	Aneroid	592	592	4,390 m. / 14,403 ft.	3·7	Overcast till 7; then cleared up.

*Calculated relatively to Ibanda.

Appendix B.

TABLE VIII—*continued.*

Number.	Place.	Corresponding Number of the Map.	Date.		Type of Barometer.	Altitude.			Temperature Centigrades.	State of the Weather.
			Month and Day.	Hour.		Above Bujongolo.		Above the sea.		
						Each separately. m.	Mean. m.			
37	Lake Bujuku ..	33	14 July	10. 30	Hypsometer	120	120	3,918 m. 12,855 ft.	10·7	From and after 15th July, the heights were obtained by comparison with the data of Ibanda (altitude, 148 metres, relatively to Fort Portal, hence, 13·4 metres, or 44¼ feet above sea-level).
38	Bujuku Valley, first terrace beyond the lake	34	14 „	..	Aneroid	− 50	50	3,748 m. 12,297 ft.	..	
39	Bujuku Valley, clearing beyond the first terrace	34 twice	14 „	..	„	− 210	210	3,588 m. 11,771 ft.	..	
40	Bujuku Valley, second terrace beyond the clearing	35	14 „	..	„	− 260	260	3,538 m. 11,608 ft.	..	
41	Camp IX	36	14 „	..	Hypsometer	− 292	292	3,506 m. 11,503 ft.	9·7	
42	Migiusi Valley, depression first plain	37	15 „	8	Aneroid	− 156	156	3,642 m. 11,949 ft.		
43	Migiusi Valley, centre..	38	15 „	10. 15	Hypsometer	− 3	3	3,795 m. 12,451 ft.	6·6	
44	Camp X..	39	15 „	..	„	368	368	4,166 m. 13,668 ft.	0·6	
45	Camp XI	40	17 „	15	„	− 888	888	2,910 m. 9,547 ft.	0·6	
46	Morena, depression below Nakitawa	41	18 „	15	„	−1802	1802	1,996 m. 6,549 ft.	16·6	

APPENDIX C.

CONTAINS A SUMMARY OF THE GEOLOGICAL, PETROGRAPHIC, AND MINERALOGICAL OBSERVATIONS WHICH WERE MADE BY H.R.H. THE DUKE OF THE ABRUZZI'S RUWENZORI EXPEDITION, TOGETHER WITH A LIST OF THE NEW ZOOLOGICAL AND BOTANICAL GENERA AND SPECIES COLLECTED IN THE RUWENZORI REGIONS.

NOTE.—All these observations are fully embodied in the scientific volume which is published only in Italian, and in which are also described and illustrated the new zoological and botanical specimens.

SUMMARY

OF THE GEOLOGICAL OBSERVATIONS MADE IN UGANDA AND IN THE RUWENZORI RANGE DURING THE EXPEDITION OF H.R.H. THE DUKE OF THE ABRUZZI,

BY

ALESSANDRO ROCCATI.

I.—UGANDA.

Overlooking for the present the recent surface formations of concretionary limonite and of laterite, that part of Uganda which was traversed by H.R.H. the Duke of the Abruzzi's Expedition was found to be for the most part constituted of the crystalline rocks which form the archæan plateau of Central Africa. A not inconsiderable tract, however, of the region traversed is covered with sedimentary formations referable to the Palæozoic Age, and in the intermediate neighbourhood of Fort Portal there is an apparently limited zone in which the crystalline rocks are overlaid by recent volcanic formations, represented by stratified tuffs which I take to be of subaqueous origin.

Archæan.—On leaving Entebbe in the direction of the west, archæan crystalline rocks are followed from the shores of Lake Victoria nearly to Mitiana. Here begin to appear the palæozoic formations, which, interrupted here and there by out-cropping *coarse-grained granite* and *pegmatite*, extend to within a few miles of Kasiba, where they suddenly disappear altogether, their place being taken by the crystalline rocks, which continue westwards without further break and thus constitute the whole of the Ruwenzori Range.

The archæan rocks are represented by *mica-schists*, *gneiss*, and *granites* (always associated with numerous quartzites), amongst which are here and there noticed intrusive *greenstones*, and seams of *pegmatite, microgranite*, etc.

Summary of Geological Observations.

All the gneiss and granitic rocks are strikingly conformable in their structure and composition, and their characters agree, broadly speaking, with the descriptions already given by those writers who have occupied themselves with the crystalline formations of South and Central Africa, thus further showing the prevailing uniformity in the constitution of the extensive archæan plateau.

In the first archæan zone, that is, between the shores of the lake and the overlying primitive formations near Mitiana, the *mica-schists* seem to prevail, these being associated with *gneiss* and thick *quartzite* beds. The mica of the micaceous schists is the *muscovite* variety, which occurs in large laminæ, thus forming rocks of a very marked schistose type, but always highly reddened, and often somewhat disintegrated by aerial denudation.

When we pass beyond the palæozoic and thus re-enter the archæan zone, we find the region between Kasiba and Muyongo constituted entirely of a *large-grained granite*, which appears to form a range running about south to north. In this granite are noticed hydiomorphic forms of *felspar*, which in their greatest development reach two inches and more. In the granular mass of the rock *quartz* abounds, while the *biotite* mica is, as a rule, relatively scarce. Throughout the whole region the granite is always profoundly metamorphized, a fact which contrasts with the relatively fresh aspect of the rock in the granitic outcrops of the palæozoic zone.

At Muyongo the mica-schists again become associated with gneiss, the latter predominating. Such association, always accompanied by *quartzite beds*, and in places by *minute biotite mica-schists*, and by *talc-schists*, is continued right up into the Ruwenzori Range, into the constituents of which it largely enters.

The gneiss is of a schistose character which is never very distinct, hence is to be considered as a *granitic gneiss*, the micaceous element of which is *biotite*, and presents an always more or less pronounced kataclastic structure. Characteristic of this rock are everywhere the really considerable abundance of the ferruginous minerals, such as *magnetite*, *ilmenite*, and *hematite* (the changes of which explain the frequent reddish surface of gneiss), and the constant presence of microcline, which becomes the prevailing, one may even say often the exclusive, felspar variety in this rock. This indeed is a fact which has already been recorded in other parts of Central and South Africa.

In the neighbourhood of Ruwenzori biotite gneiss, either normal or with a predominance of microcline, is partly replaced by *amphibolic gneisses*. In several districts, but especially in the granitic range between Kasiba and Muyongo, are noticed outcrops of *pegmatite* and *micro-granite*; here the pegmatite has never the coarse-grained structure comparable to that of granite,

383

Appendix C.

although some specimens present a typical graphic association of quartz with microcline.

In the Lwamutukuza, Muyongo and Fort Portal districts I noticed in the gneiss-granitic formation considerable intrusions of *diabase* rocks of granular and sometimes coarse-grained structure. The specimens collected by us never contain olivine, nor even the chloritic green pigment so common in the rocks of this type in our lands; characteristic is always the abundance of ilmenite, as also the basic felspar often referable to *anorthite*.

Thanks to the metamorphosis of the pyroxenes in amphiboles, which may be easily followed in its various transitions, some of these diabases pass over to *epidiorite*; true *diorite* I did not come across *in situ*, but believe that it occurs in the Kaibo-Butiti district. Conspicuous also, between Fort Portal and Duwona, is a thick bed of overlying *hypersthene gabbro* of coarse structure.

Palæozoic.—The formations which represent the Palæozoic Age follow for about 50 miles between Mitiana and Kasiba. Their eastern limit did not appear to be very clearly marked, whereas the western is distinctly defined by the granitic range which I have described as extending from Kasiba to Muyongo. It is in fact against these very escarpments that the palæozoic formations are inclined.

The rocks met in the district are *sandstones, arkoses, quartzites, quartzite breccias* and various *schists, micaceous* or *talco-micaceous.* All these rocks, whose clastic and metamorphic origin is readily recognized in the petrographic laboratory, are for the most part coloured a deep red, and correspond perfectly to the rocks referred to the Palæozoic Age, as described by observers in other parts of Uganda, as well as in Congoland and South Africa.

An exact determination of age is too often prevented by a total lack of fossils. I think, however, that it may be useful to point out how some of the schists met by me greatly resemble analogous formations of the Permian Epoch in the Alps, and how, as we proceed westwards, the series seem distinctly to pass from the sandstones to the schists, thus suggesting a steady increase of metamorphism in that direction.

Recent formations.—These are represented by the *concretionary limonite* (the ironstone of English writers), and by *laterite*.

The *concretionary limonite* is one of the characteristic formations of the Lake Victoria region.

Already on the east shore, and then in a typical manner on the west, in the Entebbe district and beyond it, we may say as far as the Kasiba-Muyongo granitic zone, the ground is covered with a concretionary limestone, at times pisolitic (pea-like) or vacuolated, always very compact, colour shifting from a bright red to a brownish-yellow or dark brown.

Summary of Geological Observations.

It supplies the building material adopted for the structures of European type at Entebbe, Mitiana, and other places. Its chemical composition is very constant, containing a percentage of Fe 20_3, which varies from 51 to 55 per cent. ; hence it differs from that of the limonite nodules, which are found in laterite, in which the percentage of Fe 20_3 may be as high as 82 per cent. In this region the limonite forms several rising grounds, some over 300 feet above the present level of the lake. As to its origin, I believe it was deposited in the bed of the lake, at a time when it was far more extensive than at present, as is evident from a whole series of indisputable facts, the decomposition being effected by a mechanical process analogous to that which in lacustrine basins originates the limonite (bog ore) of marshy places.

Hence, in my opinion, the concretionary limonite may be of great importance, as serving to indicate the former limits reached by Lake Victoria, of which even Lake Isolt, near Bujongo, may be merely a remnant. Similarly the few patches of concretionary limonite met in the Butiti district may possibly represent old extensions of Lake Albert. In the limonite are embedded numerous stony fragments and nodules varying greatly in size. This detrital material is for the most part represented by quartz, hyaline (glassy), granular, or jasproid. I rarely noticed nodules or fragments of gneiss, or of palæozic rocks. So great at times is the abundance of quartzose nodules or fragments as to give rise either to conglomerates or else to breccias with limonitic cement, as the case may be.

The scenery of the limonite region is typical in the form of its rising grounds. These do not present rounded contours, such as are normally observed in other parts of Uganda, but constitute elongated hills with levelled summits, divided one from another by deep fissures, or else they stand isolated on the plain, representing the remains of what at other times must have been the unbroken surface of the ground. In this respect the Entebbe and Mitiana districts are characteristic.

The *laterite* resulting from the transformation of the felspar rocks, under the action of the atmospheric agencies, aided by the high temperature and by the alternating droughts and heavy rains, may be said to form the surface layer of the ground throughout this region. It is found not only in the gneiss and granite zones, but also where the palæozoic occurs and forms on the rocks *in situ* a covering of various thickness, which may in places acquire quite an exceptional development. Characteristic is always the inner red colour, which is due to the excessive oxidation of the numerous iron ores that we have seen to be present in those rocks.

In the laterite is often noticed an abundant *micaceous hematite*, which in

Appendix C.

some places is accumulated by the rain waters in depressions of the ground. Common also at varying depths are great beds of *limonite*, which in some places, as at Butiti, are accompanied by *oxides* of *manganese*.

This limonite is mined by the natives, who, by the Catalonian process, extract from it the iron which they use in the manufacture of arms and implements, an industry in which they display much skill.

Physical Features.—Owing to the abundant herbaceous vegetation, and to the thick deposits of laterite and concretionary limonite, which everywhere cover the ground, I was unable to make any conclusive stratigraphic observations. Nevertheless, from the data which we collected it seems quite evident that there must be a considerable discordance between the archæan and palæozoic formations.

Erosion.—Amongst the phenomena of erosion, which, as may be easily understood, are very marked, mention should be made of the denudation, thanks to which the older rocks, being deprived of their laterite covering, become exposed on the surface of the ground. They usually assume the characteristic aspect of mammiform or hummocky rocks, the so-called *roches moutonnées* of the French, which so strangely resemble the glacial features of our lands. The fact is explained by the absence of the factor of frost and thaw, in consequence of which the rocks, instead of becoming disintegrated, suffer only a surface change and rupture, the rubbly fragments of which get constantly displaced and washed away by the rain waters.

Outwardly the rocks often present a crust of varying thickness, which is due to metamorphism, and this crust adheres in the loosest way to the underlying mass. Not seldom the adherence fails altogether, and then the transformed surface forms slabs with rounded edges merely resting on the underlying rock, which is still relatively intact.

This phenomenon is seen where the gneisses crop out. In the case of granites there is further noticed a cleavage of the rock in great blocks of parallelopiped form which, presenting greater resistance to decomposition, end by being at last completely isolated and detached. In the vicinity of Muyongo hundreds of such masses occur in the form of prisms, cubes and obelisks, at times of remarkably regular outline.

Another consequence of this predominantly superficial disintegration is a peculiar ruggedness which is presented by the surface of the rocks, and is due to the protruding quartz that resists the decomposing forces, while the felspar part is broken up and carried away by the water. In the zone of the coarse-grained granite this protrusion on the surface is noticed even in the case of the largely hydiomorphic crystals of felspar.

Summary of Geological Observations.

In some districts, as at Kaibo and in the neighbourhood of Fort Portal, are found isolated masses, or accumulations of masses, on the summits or the flanks of hilly elevations consisting of laterite. In the specified districts these masses consist of diabase, and we may take it that their presence is due to the resistance of certain rocks, perhaps originally in the form of dikes, and to the metamorphism which reduced to laterite the gneiss or granite in which they were embedded. In some places it is not improbable that it may be a question of some kind of transport.

Vulcanism.—Recent igneous formations are met at the eastern foot of Ruwenzori. Here they serve to indicate the presence of one or more lines of fracture in relation with that great Rift Valley with which originated the depression comprising Lakes Tanganika, Kivu, Albert Edward, Albert, and the Semliki Valley, and which contributed to the isolation of the Ruwenzori Range. In the Fort Portal district volcanic action is indicated by thermal springs (Butanuka), and by stratified tuff which cover the ground and form a series of little volcanoes, whose craters are now mostly flooded with tarns. They form a chain which is disposed very nearly in the direction from south to north.

The tuffs of this formation are partly compact and partly of loose structure. All, however, are of subaqueous origin, and thus attest the greater extension in former times occupied by Lake Albert Edward, which must probably have been united with Lake Albert towards the north.

The compact tuffs are of a dark hue, and very hard, and yield a cement consisting of a basic silicate rich in iron and easily decomposed by acids. Such tuffs occur in all the craters of the series, only more or less transformed, the change consisting in a tendency to acquire a red colour due to the decomposition of the silicate of iron.

The tuffs of looser structure, which are met partly in the craters and in all the surface formations, are of a colour passing from white to grey. They derive principally from fragments of the compact tuff cemented by calcite after the complete discoloration caused by the metamorphic process. The enclosed exotic fragments are numerous, especially in the non-compact variety, and they consist of fragments of the most diverse sizes, whether water-borne or not. These ingredients may for the most part be considered as coming from rocks of the Ruwenzori Range, such as gneisses, diabases, diorites, garnet-bearing rocks, amphibolites, etc. In the hill at Fort Portal the tuffs are, moreover, rich in vegetable remains which, unfortunately, cannot now be determined. In the Butiti—Fort Portal region, mineral springs are also numerous, and the country is subject to frequent earthquakes.

Appendix C.

II.—RUWENZORI RANGE.

Lithological Constitution.—The Ruwenzori Range, as already pointed out by Scott Elliot, is regarded as a part of the archæan formation of Equatorial Africa which has been upheaved through phenomena of dislocation. In fact, it is found to be essentially constituted of various *gneisses* and *mica-schists* in which must have been originally embedded the *greenstones* which, after being exposed by the phenomena of denudation, to which they offered the greatest resistance, now form the loftiest crests of the whole range. Ascending the valley of the Mobuku, we meet with a regular succession of rocks, in which first occur the *gneisses* whose correspondence with the analogous rocks of Uganda is obvious.

Amongst these gneisses the dominant variety appears to be the *biotite* and *microcline* of the Fort Portal district. With it are associated *micaceous-amphibolitics* and amphibolitic varieties in which the *amphibole* is represented by *hornblende*. In the amphibolic gneisses, however, the microcline is rare or absent, being displaced, besides *orthoclase*, by an abundance of *plagioclase* referable to *andesite*.

The *kataclastic* structure still continues together with frequent metallic ores, such as *hematite, magnetite, ilmenite,* and *chromite,* some varieties being rich in *tourmaline* and *garnet.*

The gneisses range up to about 11,600 feet in association with *mica-schists,* the first type, however, being always dominant. *Quartzites* also abound both in thick beds and nodules.

Above 11,600 feet the gneisses disappear, the *mica-schists* alone persisting, associated with *quartzites,* and following without break up to the zone of the *greenstones.*

The mica-schists of the Mobuku Valley are of two kinds, which constantly recur : *minute* and *foliaceous* (*lamellar*).

The first are formed of minute *muscovite laminæ*, with abundant quartz, and next to it *felspar*, mostly *andesite ;* in these the schistosity is not always evident, while their compactness and hardness are very great.

In the foliaceous kind *muscovite* prevails in large white silvery laminæ, with which is associated a little minute *biotite*, while *quartz* and *felspar* become rare. In this second variety the schistosity and the cleavage are clearly seen.

The two mica-schist types form beds of varying thickness, either standing quite apart or else passing gradually into one another. They are always and

Summary of Geological Observations.

everywhere rich in *tourmaline* and *metallic* ores, *ilmenite, chromite, hematite* and *magnetite*. In some places *garnet* and *apatite* are also noticed, while in the schistose surfaces fine fibrous aggregates of *cyanite* and *sillimanite* are common.

On the Kichuchu Plain, besides the existence of a *labradorite gneiss*, the presence is conspicuous of some dikes of *basalt*, which ramify and intersect the gneiss-mica-schist formation. This basalt, which constitutes the only evidence of recent volcanic action met by us in the range, is microcrystalline and of holocrystalline type. On the Biamba Plain I further met a *diabase* in which the opaque element is represented by *chromite* alone.

Towards 12,000 feet the zone of the mica-schists disappears, and the greenstones come to the surface; these constitute exclusively Mts. Baker and Stanley, and are associated with gneiss on the other heights visited by the expedition. On the western slope of Mt. Baker the identical mica-schists reappear, which we had met in the Mobuku Valley, but on the west side they range somewhat higher than on the east.

The zone of the greenstones is constituted essentially of an *amphibolite schist*, in which the schistose element is more or less evident; it is usually micro-crystalline and formed of *hornblende* with quartz, and in the second place *felspar* (mostly andesite), and in this case it passes over to a *diorite schist*. From this amphibolic schist are developed some varieties due to the substitution of *actinolite* for *hornblende*, or else to its association with *garnet, biotite* and *pyroxene*.

Abundant in these rocks are *ilmenite* and *epidote*, the latter also forming numerous beds, veins and nodules, some of which are of extraordinary thickness, as much as 30 feet in the longer axis. Moreover, numerous beds of *quartzite* everywhere accompany the amphibolic schists, with which in the various mountains are associated other rocks in the following way :—

MT. BAKER.—*Quartziferous diorite; compact amphibolite* which forms the Edward Peak, on the summit of which are numerous fulgurites; *crystalline limestone; chlorite schist, epidosyte, grenatite; diabase.*

At several points on this mountain are noticed some lenticels, geodes and small veins of *pyrites, calco-pyrites* and *ilmenite*, with *felspars, quartz* and *calcite;* on Wollaston Peak a small vein of *galena* with a gangue of calcite crops out.

MT. STANLEY.—*Compact amphibolite; amphibolic schist* with large *garnets, diorite* and *labradorite diorite* which forms the Alexandra Peak, and probably also the Margherita; it is noted for its various types of *fulgurites, epidosites* and *diabase.*

On this mountain also *pyrite* and especially *ilmenite* are plentiful, as are also copper ores : *chalco-pyrite tetrahedrites, malachite.*

MT. LUIGI DI SAVOIA.—Here also amphibolic schist crops out, although

389

Appendix C.

the mountain is essentially constituted of gneiss, the *biotite* variety and *microcline* being common on the lower parts of the mountain. This gneiss, associated with *mica-schists*, may be traced from Ibanda by the Mahoma Valley, not only up to the crests of the mountain, but probably also extends to the south and west of the range.

It should be mentioned that in Mt. Luigi di Savoia there are large dikes of macroscopic *pegmatite* rich in *garnet* and *tourmaline*, haplite and micro-granite in the neighbourhood of Stairs Peak ; *diabase*, which crops out at Sella Peak, where it abounds in fulgurites ; *diorite, epidosyte* and crystalline chalk, which seems to point at contact between the gneisses and the amphibolic schists.

MT. SPEKE.—The prevailing rock appears to be a granitoid gneiss with biotite and abundant *epidote ;* with the gneiss would appear to be associated *diorite, amphibolite* and micro-granite.

MT. EMIN.—Yields *quartzite* and a *diorite* analogous to that of Mt. Stanley.

MT. GESSI.—The dominant rock again appears to be *amphibolic* schist in association with *quartzite* and *epidosyte.*

In the Bujuku Valley the prevalent form appears to be of a type analogous to that occurring on Mt. Speke. This valley, as well as that of the Mubuku, would seem in its upper reaches to open out in contact with gneiss and amphibolic rocks.

Tectonic Structure.—The tectonic feature by which the Ruwenzori Massif is outlined and clearly characterized is represented by two great zones of fracture. One lying to the west is of vast size, having given rise to the Semliki Valley, and in this direction completely isolated the enormous mass of the Ruwenzori Range. The other (eastern) zone of fracture is less marked, but well outlined by the volcanic formations, in which are included those of Fort Portal.

In relation with the two main zones of fracture, others occur in the interior of the range, and these are disposed in two different directions, one west and east—that is to say, normal to the chief trends—the other, on the contrary, running in parallel lines from south to north. To these lines of inner fracture are due several valleys and many of the secondary glens, which tend to give their characteristic isolation to all the principal heights.

The stratigraphic disposition is regular. As we ascend the Mobuku Valley, we everywhere notice in the gneiss and mica-schist beds an incline from east to south-east. This incline is, on the whole, maintained in Mt. Baker, and is clearly seen, for instance, in Cagni Peak. In Mt. Luigi di Savoia the east-south-east slope recurs, with a tendency to the south which farther on becomes due south. In Mt. Stanley the south-east tends to change to west or north-west, although the south-east to east incline reappears in the Bujuku

390

Summary of Geological Observations.

Valley. Moreover, the slopes of the strata are everywhere very steep, in places as much as and upwards of 60°.

Ruwenzori must accordingly be regarded as resulting from an anticlinal or ellipsoidic upheaval, with a slope to the west on the west side, to the east on the east side, passing to the south on the south side, and probably to the north on the north side.

The presence of this ellipsoidic upheaval, combined with the phenomenon of the great fractures above mentioned, and with the existence in the central parts of rocks resisting subaerial disintegration, would explain the origin of the Ruwenzori Range and of its lofty summits.

Old Glaciation.—A phenomenon of great importance is the vast development of the glaciers of the Ruwenzori Range during the glacial period.

The valleys of the Mobuku, the Bujuku, and the Mahoma were filled by the glaciers which descended from the chief mountains. These uniting in a single ice-stream of great size, and filling up the Lower Mobuku Valley, must have easily extended as far as the plain of Ibanda.

Proofs of this early glacial expansion are afforded by the numerous large erratic blocks; by the old moraines which occupy the Mobuku Valley from Bihunga to Kichuchu, and above which rises the Nakitawa Plain; lastly, by the rolled and striated rocks which are so common on the higher parts of the mountain. Regarding the Nakitawa moraine, it may be mentioned that the lake lying south-west of that district, and by the observers generally considered as volcanic, is, on the contrary, inter-morainic.

Disregarding the erratic boulders which occur on the plain of Ibanda, and are not perhaps due to direct glacial transportation, the first undoubted proofs of old glaciation were met by me near the ascent of Bihunga, that is, at about 4,500 feet, whereas at present the glaciers do not descend lower than about 12,600 feet.

On the western slope, too, the traces are evident of the passage of the old glaciers with scratched and rounded blocks and morainic formations. We were, however, unable to discover how far they had ranged on that side, as we did not advance very far in that direction.

Recent Glaciation.—The Ruwenzori glaciers are referred to the so-called *equatorial* type; that is to say, they form ice-caps which are at times of great thickness, and more or less completely cover the summits of the mountains. From these ice-caps branches ramify downwards and advance into the ravines, but seldom range, and then only a little way, beyond the lower level of the perennial snows, which here lies between 13,350 and 13,500 feet.

The position of the glaciers once determined, the lateral moraines may be

391

Appendix C.

neglected; nor do the underlying ones appear to have any great developments, judging at least from the frontal moraines, which are never very extensive.

The position of the glaciers likewise includes the existence of depressions in which snow might be collected; falling on the whole surface of the glacier, the snow passes directly and rapidly to the state of ice, a phenomenon which is easily explained by the atmospheric conditions of these highlands, which, during certain hours of the day often tend to develop high temperatures.

One of the characteristics of the Ruwenzori glaciers is the presence of enormous cornices from which hang multitudes of large stalactites, which become a firm support to the cornices themselves. The origin of these curious stalactites is again to be sought in the special meteorological conditions, which tend to rapid changes of temperature not only between day and night, but also at different times of the day itself, according to the state of the weather.

Another noteworthy feature is the water welling up in front of the glaciers, which never presents that turbid look which, under like conditions, is seen in the melting waters of the Alpine glaciers. The water is perfectly limpid, which shows that the movement of the glaciers is but slight, at least at present. Hence the erosion must also be insignificant, and this again explains the absence of considerable underlying moraines.

In fact, all the Ruwenzori glaciers are nowadays in a state of rapid retreat. Of this proof is afforded in the recently abandoned morainic formations which are noticed in many places; in the wide areas of polished rocks at the sides and in front of the glaciers; in a zone not yet invaded by the mosses and lichens, which are typically abundant even on the most elevated tracts of the range; lastly, in the whitish colour so often noticed on the surface of those rocks which have only recently got rid of the mantle of snow and ice by which they were formerly covered.

Erosive Phenomena.—On the lower part of the Ruwenzori Range identical climatic conditions lead to the identical phenomena of meteoric denudation that are also observed in Uganda. We have accordingly an abundant *laterite* formation on which a rank herbaceous vegetation is developed; here also is that rounded form of the exposed rocks with their sham aspect of *roches moutonnées* above indicated; further, the outward protrusion of some of the more durable components; the cleavage of the rock in superficial slabs, and so on.

The zone of the old morainic formations is clothed with a dense arboreal or bushy vegetation of tropical type, and this protects the underlying soil from erosive action. Here and there, however, are noticed rents and rifts caused by the torrential and swelling waters, with formations in some places of typical fungi-form rocks, as near Nakitawa.

Summary of Geological Observations.

Towards 9,000 feet the persistently humid climate gives rise to the zone of the cryptogams, and to the bogs that constitute one of the characteristic features of Ruwenzori. From this altitude, we may say right up to the glaciers, the ground is everywhere uninterruptedly covered with a boggy peat-turf layer which not seldom reaches or exceeds a thickness of 20 inches. On this substratum is developed a vigorous vegetation of mosses, hepaticæ (liverworts) and lichens, which spread a thick mantle over the protruding rocks, the erratic boulders and the trunks of the trees, whether living or fallen with age, and for centuries accumulating on the surface of the ground.

Over this overlying stratum of bog and vegetable detritus there is but a slight flow of water, absorbed as it is as by a huge sponge. The surface layer thus forms a protecting carpet for the rocks which, when they can be seen underneath, appear to be relatively intact, escaping as they do in great measure from the erosive phenomena.

Beyond the boggy zone the surface action of the meteoric agents comes again into play, but it must act very slowly in consequence of the abundant vegetation of the crustaceous lichens covering the rocks. The nature of these rocks, largely constituted of amphibole and quartz, also explains the slighter action of atmospheric denudation which has freer play in the gneiss and mica-schist zone.

Attention may again be called to the characteristic protrusion of the rocks composed of more resisting elements. This fact is apparent in the beds of garnet-bearing rocks, where the large crystals of garnet protrude with an almost variolar or pitted aspect. The phenomenon is typical also in the mica-schist zone on the western slope of Mt. Baker, where the mica-schist is associated with abundant quartz in lenticular veins and thin layers which everywhere form protuberances, sometimes even very conspicuous on the surface of the ground.

Lastly, in the higher zones to the modifying and erosive action of the atmosphere is added the extremely potent factor of frost and thaw. In this case, wherever the underlying rocks are not protected by the masses of ice, we find long stretches of ground covered with loose chaotic and shifting detritus analogous to what is noticed on the crests and higher slopes of our Alpine heights.

Appendix C.

ALPHABETIC LIST OF THE MINERALS COLLECTED IN THE RUWENZORI RANGE.

Actinolite.

Albite.

Apatite.

Calcite.

Chalcopyrite.

Chlorite.

Chromite.

Diopside.

Epidote.

Galena.

Garnet.

Ilmenite.

Magnetite.

Malachite.

Microcline.

Muscovite.

Pyrite.

Quartz.

Tetrahedrite.

Tourmaline.

Tremolite.

ZOOLOGY.

New Genera, Species, and Sub-species collected by the Expedition of H.R.H. the Duke of the Abruzzi.

Mammals ... Nyctinomus Aloysii Sabaudiæ, *Festa*.

Felis pardus sub. spec. Ruwenzorii, *Camerano*.

Birds ... Anthoscopus Roccatii, *Salvadori*.

Lagonosticta Ugandæ, *Salvadori*.

Bycanistes Aloysii, *Salvadori*.

Xylobucco Aloysii, *Salvadori*.

Reptiles ... Lygosoma Aloysii Sabaudiæ, *Peracca*.

Molluscs ... Ennea Roccatii, *Pollonera*.

Ennea Sellæ, *Pollonera*.

Ennea Camerani, *Pollonera*.

Ennea Aloysii Sabaudiæ, *Pollonera*.

Streptaxis Cavallii, *Pollonera*.

Urocyclus zonatus, *Pollonera*.

Zoological List.

Molluscs
—contd.

Urocyclus tenuizonatus, *Pollonera.*
Urocyclus subfasciatus, *Pollonera.*
Urocyclus raripunctatus, *Pollonera.*
Atoxon ornatum, *Pollonera.*
Atoxon Cavallii, *Pollonera.*
Dendrolimax leprosus, *Pollonera.*
Microcyclus modestus, *Pollonera.*
Microcyclus incertus, *Pollonera.*
Trichotoxon Roccatii, *Pollonera.*
Kirkia nov. gen., *Pollonera.*
Helicarion Aloysii Sabaudiæ, *Pollonera.*
Vitrina Cagnii, *Pollonera.*
Vitrina ibandensis, *Pollonera.*
Martensia entebbena, *Pollonera.*
Fruticicola bujungolensis, *Pollonera.*
Fruticicola Bihungæ, *Pollonera.*
Buliminus Aloysii Sabaudiæ, *Pollonera.*
Limicolaria tussiformis var. nov. ugandensis, *Pollonera.*
Limicolaria Roccatii, *Pollonera.*
Limicolaria rectistrigata var. nov. simplicissimus, *Pollonera*, and
    var. nov. simplex, *Pollonera.*
Limicolaria pura, *Pollonera.*
Limicolaria pura var. diluta, *Pollonera.*
Limicolaria Cavallii, *Pollonera.*
Glessula De-Albertisi, *Pollonera.*
Glessula ferussacioides, *Pollonera.*
Homorus olivaceus, *Pollonera.*
Subulina Roccatii, *Pollonera.*
Subulina Ruwenzorensis, *Pollonera.*
Subulina Ruwenzorensis var. elongata, *Pollonera.*
Vaginula Roccatii, *Pollonera.*

Beetles ...

Hydaticus Rochei, *Camerano.*
Cillæus Cavallii, *Camerano.*
Cillæus Cagnii, *Camerano.*
Hydrophilus Loanei, *Camerano.*
Lixus Roccatii, *Camerano.*
Sipalus Aloysii Sabaudiæ, *Camerano.*
Eumelosomus Aloysii Sabaudiæ, *Pangella.*

Appendix C.

Dermaptera... Pygidicrana livida, *Borelli*.
Anisolabis compressa, *Borelli*.
Genolabis picea, *Borelli*.
Spongiphora Aloysii Sabaudiæ, *Borelli*.
Chætospania ugandana, *Borelli*.
Opisthocosmia Roccatii, *Borelli*.
Apterygida Cagnii, *Borelli*.
Apterygida Cavallii, *Borelli*.

Orthoptera ... Ceratinoptera portalensis, *Giglio-Tos*.
Hemithyrsocera sabauda, *Giglio-Tos*.
Blatta ugandana, *Giglio-Tos*.
Pyrgophyma nov. gen., *Giglio-Tos*.
Pyrgophyma sabaudum, *Giglio-Tos*.
Euprepocnemis ibandana, *Giglio-Tos*.
Tylopsis dubia, *Giglio-Tos*.

Myriapoda ... Cryptops Aloysii Sabaudiæ, *Silvestri*.
Scutigerella Ruwenzorii, *Silvestri*.
Phœodesmus Aloysii Sabaudiæ, *Silvestri*.
Habrodesmus Cagnii, *Silvestri*.
Julidesmus Cavallii, *Silvestri*.
Scaptodesmus Roccatii, *Silvestri*.
Scaptodesmus rugifer, *Silvestri*.
Compsodesmus Sellæ, *Silvestri*.
Tymbodesmus insignitus, *Silvestri*.
Archispirostreptus ibanda, *Silvestri*.
Archispirostreptus virgator, *Silvestri*.
Archispirostreptus nakitawa, *Silvestri*.
Odontopyge Aloysii Sabaudiæ, *Silvestri*.
Odontopyge Winspearei, *Silvestri*.
Odontopyge Petigaxi, *Silvestri*.
Odontopyge Ollieri, *Silvestri*.

Crustacea ... Potamon Aloysii Sabaudiæ, *Nobili*.
Synarmadilloides nov. gen., *Nobili*.
Synarmadilloides Roccatii, *Nobili*.

Worms ... Dichogaster Aloysii Sabaudiæ, *Cognetti*.
Dichogaster Roccatii, *Cognetti*.
Dichogaster Cagnii, *Cognetti*.
Dichogaster excelsa, *Cognetti*.

Botanical List.

Worms
—contd.

Dichogaster Duwoni, *Cognetti.*
Dichogaster Sellæ, *Cognetti.*
Dichogaster Ruwenzorii, *Cognetti.*
Dichogaster demoniaca, *Cognetti.*
Dichogaster toroensis, *Cognetti.*
Gordiodrilus mobuccanus, *Cognetti.*
Pareudrilus pallidus, *Cognetti.*
Eminoscolex Rochei, *Cognetti.*
Eminoscolex Nakitavæ, *Cognetti.*
Neumanniella æquatorialis, *Cognetti.*
Alma Aloysii Sabaudiæ, *Cognetti.*

Nematoids ...

Strongylus minutoides, *Parona.*
Strongylus Cavallii, *Parona.*
Uncinaria muridis, *Parona.*
Physaloptera Aloysii Sabaudiæ, *Parona.*
Physaloptera Ruwenzorii, *Parona.*

Of all the groups of animals above-mentioned the expedition collected other already known species ; many of these had not yet been recorded in the Uganda and Ruwenzori regions. Hence, in respect of the distribution of animal species also, the expedition has made valuable contributions to our knowledge of the African fauna.

SUMMARY OF THE PLANTS COLLECTED BY THE EXPEDITION OF THE DUKE OF THE ABRUZZI ON THE RUWENZORI RANGE.

Embryophyta Siphonagama (Auct. E. Chiovenda et F. Cortesi).

Species collected ... ... 93.
New Species ... ... 18.

Graminaceæ ...

1. Andropogon mobukensis, *Chiov.*
2. Deschampsia ruwensorensis, *Chiov.*
3. Festuca gelida, *Chiov.*
4. Oxytenanthera ? ruwensorensis, *Chiov.*

397

Appendix C.

Asteraceæ ... 5. Helichrysum Ducis Aprutii, *Chiov*
6. Senecio coreopsoides, *Chiov.*
7. Senecio Pirottæ, *Chiov.*
8. Senecio Mattirolii, *Chiov.*
9. Senecio Ducis Aprutii, *Chiov.*
10. Senecio Roccatii, *Chiov.*
11. Carduus blepharoleptis, *Chiov.*
12. Erlangea squarrosula, *Chiov.*

Rosaceæ ... 13. Alchemilla Roccatii, *Cort.*
14. Alchemilla Ducis Aprutii, *Cort.*
15. Alchemilla tridentata, *Cort.*

Rubiaceæ ... 16. Rubia ruwenzorensis, *Cort.*

Urticaceæ ... 17. Parietaria ruwenzorensis, *Cort.*

Crassulaceæ ... 18. Sedum Ducis Aprutii, *Cort.*

Pteridophyta (Auct. R. Pirotta).

Species collected : Hymenophyllaceæ... ... 1
Cyatheaceæ 1
Polypodiaceæ 20
Lycopodiaceæ 2
————
. 24

New species ... 4

Cyatheaceæ ... ... Cyathea Sellæ, *Pirotta* (ad.) (int.).

Polypodiaceæ ... Woodsia nivalis, *Pirotta.*
Asplenium Ducis Aprutii, *Pirotta.*
Elaphoglossum Ruwenzorii, *Pirotta.*

Musci (Auct. G. Negri).

Species collected ... 38
New species 22
Sphagnum Aloysii Sabaudiæ, *Negri.*
Sphagnum Ruwenzorense, *Negri.*
Dicranum petrophilum, *Negri.*
Campylopus sericeous, *Negri.*
Campylopus Cagnii, *Negri.*
Fissidens Mobukensis, *Negri.*

Botanical List.

Musci—*contd.*

Leptdontium Gambaragaræ, *Negri.*
Tortula Cavallii, *Negri.*
Anoectangium Sellæ, *Negri.*
Anoectangium fuscum, *Negri.*
Anoectangium flexuosum, *Negri.*
Zygodon Roccatii, *Negri.*
Zygodon hirsutum, *Negri.*
Amphydium Aloysii Sabaudiæ, *Negri.*
Macromitrium fragile, *Negri.*
Brachymenium Cagnii, *Negri.*
Pohlia Aloysii Sabaudiæ, *Negri.*
Bryum Sellæ, *Negri.*
Breutelia auronitens, *Negri.*
Catharinæa Cavallii, *Negri.*
Polytrichum cupreum, *Negri.*
Brachythecium Roccatii, *Negri.*

Hepaticæ (Auct. G. Gola).

Species collected : Marchantiaceæ sp.   3
Jungermanniaceæ anakrogynæ sp. ...   4
Jungermanniaceæ akrogynæ sp.   ...   26

sp. 33

New species   ...   16

Marchantia Cagnii, *Gola.*
Marchantia Sellæ, *Gola.*
Marchantia papyracæ, *Gola.*
Metzgeria ruwenzorensis, *Gola.*
Symphogyna Sellæ, *Gola.*
Symphogyna Aloysii Sabaudiæ, *Gola.*
Anastrophyllum Gambaragaræ, *Gola.*
Plagiochila lævifolia, *Gola.*
Plagiochila Aloysii Sabaudiæ, *Gola.*
Lophocolea Cagnii, *Gola.*
Bazzania Roccatii, *Gola.*
Blepharostomum Cavallii, *Gola.*
Microlejeunea magnilobula, *Gola.*
Acrolejeunea fuscescens, *Gola.*

399

Appendix C.

Hepaticæ—*contd.*

Acrolejeunea Roccatii, *Gola.*
Frullania Cavallii, *Gola.*

Lichenes (Auct. A. Jatta).

Species collected 83
New species   ... 5 (var. 4)

Usnea arthroclada Fee v. ruvidescens, *Jatta.*
Parmelia Ducalis, *Jatta.*
Anaptychia leucomela Tre. v. soredica, *Jatta.*
Caloplaca citrinella, *Jatta.*
Pertusaria Roccatii, *Jatta.*
Phlyctis Ruwenzorensis, *Jatta.*
Cladonia squamosa Hffm. v. macra, *Jatta.*
Gyrophora haplocarpa Nyl. v. africana, *Jatta.*
Lecidea Cagnii, *Jatta.*

Algæ (Auct. G. B. Detoni et A. Forti).

Species collected : Myxophyceæ   ...2 et var. 1
Clorophyceæ   2
Bacillariaceæ   35 et 34 var. et form

New varieties :   Navicula borealis, *Kuetz.*
Var. exilis, *Detoni et Forti.*
Suriraya ovalis, *Breb.*
Var. enormis, *Detoni et Forti.*

Fungi (O. Mattirolo).

Species collected 27

New genus : Aloysiella, *Mattirolo et Saccardo.*

New species : Chætomella Cavalli, *Mattirolo* (Sphæropsideæ).
Aloysiella ruwenzorensis, *Mattirolo et Saccardo* (Sphæriales).
Hypoxylon crassum, *Mattirolo et Saccardo* (Sphæriales).
Cladoderris Roccati, *Mattirolo* (Thelephoreæ).
Favolaschia Cagni, *Mattirolo* (Polyporeæ).
Psylocybe Sellæ, *Mattirolo et Bresadola* (Agaricineæ).

Botanical List.

SUMMARY.

	Total of the collected Species.	Total of the Varieties.	New Genera.	New Species.	New Varieties.
Embryophyta siphonogama ...	93	...	...	18	...
Pteridophyta	24	...	...	4	...
Musci	38	...	...	22	...
Hepaticæ	33	...	...	16	...
Lichenes	83	...	...	5	4
Algæ	39	35	15	...	2
Fungi	27	...	1	6	...
	337	35	16	71	6

INDEX.

Index.

Index.

Index.